DATE DUE

Inc. 38-293

Practical Astronomy

Springer

London
Berlin
Heidelberg
New York
Barcelona
Budapest
Hong Kong
Milan
Paris
Santa Clara
Singapore
Tokyo

Other titles in this series

The Modern Amateur Astronomer
Patrick Moore (Ed.)

Telescopes and Techniques: An Introduction to
Practical Astronomy
C. R. Kitchin

The Observational Amateur Astronomer

Patrick Moore (Ed.)

Springer

Cover illustration: The constellation Orion (John Sanford/Science Photo Library).

ISBN 3-540-19899-7 Springer-Verlag Berlin Heidelberg New York
ISBN 0-387-19899-7 Springer-Verlag New York Berlin Heidelberg

British Library Cataloguing in Publication Data
Observational Amateur Astronomer. –
(Practical Astronomy Series)
I. Moore, Patrick   II. Series
520
ISBN 3-540-19899-7

Library of Congress Cataloging-in-Publication Data
The observational amateur astronomer / Patrick Moore (ed.).
      p.   cm. -- (Practical astronomy)
Includes index.
ISBN 3-540-19899-7 (pbk. : alk. paper)
1. Astronomy--Amateurs' manuals.   I. Moore, Patrick.   II. Series.
QB63.O27   1995   95-32988
520--dc20   CIP

© Springer-Verlag London Limited 1995
Printed in Great Britain

Typeset by Editburo, Lewes, East Sussex, England
Printed by the Alden Press Ltd., Osney Mead, Oxford
34/3830-543210 Printed on acid-free paper

Contents

Contents

Introduction

Amateur observers have always played a major rôle in astronomy. This is true even today, when techniques have become so advanced and so sophisticated. In *The Modern Amateur Astronomer*, instruments and techniques were discussed. The present volume is devoted entirely to observation.

Only a few decades ago, work of this kind was limited mainly to a few restricted fields: lunar and planetary observation, comet-hunting, meteor-watching, and variable star estimates. The equipment used consisted of a simple telescope of modest aperture, together with a camera which would nowadays be regarded as primitive. This is not to belittle what was achieved; far from it – and it is worth remembering that, until almost the start of the space age, the best and most detailed maps of the Moon, for example, were of amateur construction. But now the amateur has had to become more specialised if he wants to undertake really useful research, and there is a dearth of books which bridge the gap between the beginner and the really experienced worker.

This is the aim of this book. A limited amount of prior knowledge is assumed, but there is nothing here which will puzzle anyone who has done a certain amount of reading from the more elementary books on astronomy.

Each chapter has been written by an experienced observer who has specialised knowledge of the subject under discussion – and it is assumed that the equipment needed is of the type which is available to most enthusiasts at reasonable cost. We have made no attempt to impose a 'standard style' on any of the chapters, and readers will discover a variety of approaches to the subject;

there is still room for individualism in astronomy.

Amateur work today is as valuable as ever, and to a considerable extent professional researchers depend upon it. Certainly there is a degree of cooperation in astronomy not found in most other sciences; and despite the great modern observatories and the space missions, there remains plenty for the dedicated amateur to contribute.

I hope that this book will be of real use.

Patrick Moore

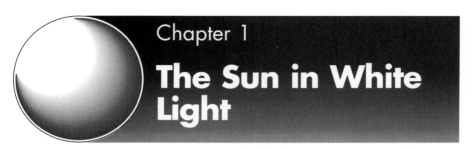

Chapter 1

The Sun in White Light

Bruce Hardie

The Sun is the only body that is dangerous to observe. Without proper protection, even a glimpse at it through a telescope or binoculars can burn the eye's retina and lead to permanent blindness.

There are two ways to observe the Sun safely: by direct viewing with a proper, fitted filter over the front of the telescope, and by projection of the Sun's image on to a card.

Direct Viewing

Safe, metallised solar filters are available. These protect the eye against both visible and invisible radiation, and protect the telescope itself against heat. They are often known as 'aperture filters', and have made every other light-reducing device obsolete as far as most amateurs are concerned.

Aperture filters come in two kinds. The first is made of metallised Mylar plastic, which usually makes the Sun look blue. The second is a metal-on-glass filter which leaves the Sun with a more natural tint: it is more durable than Mylar, but costs a lot more. To help reduce the incoming light, but mostly because of the very high cost of making a large metal-on-glass filter, the filter often incorporates an aperture stop, i.e. it has a smaller diameter than the primary. For both types, it is important that the filters should have both sides of the Mylar or glass metallised. This keeps the inevitable tiny scratches and pinholes that occur in a single-sided coating from letting sunlight through, where it will reduce image contrast by diffraction and in some cases even threaten to damage the observer's eye.

As a precaution, hold the filter up to the Sun before using it. If bright pinpoints show through they should be touched out with opaque paint. If the flaws are many and large, the filter must be discarded.

It is also vital to make sure that the filter is attached securely to the front of the telescope, so that wind or a careless knock cannot dislodge it while you are observing the Sun.

After checking and fitting the aperture filter, the first task is to orient the Sun's image. The telescope should be equatorially mounted and accurately lined up before making this observation. It should be capable of tracking without continual adjustment. A cross-wire eyepiece is practically essential; the magnification of the eyepiece should allow for the entire disc of the Sun to be seen at one time.

First we must determine the apparent directions of East and West in the sky. With the telescope stationary, rotate the cross-wire with the eyepiece so that a sunspot moves parallel to the cross-line as the Sun drifts through the field of view. Because of the Earth's rotation, the Sun's image will always appear to move through the field in a direction which is from East to West. As it drifts through the field of view the Sun's western limb will enter and disappear first, followed by the eastern limb. During use, the cross of the graticule is set centrally on the solar image and is rotated until one of the cross-lines coincides with the E–W drift of the Sun across the field of view.

To assess the position of sunspots, the graticule cross is held on the centre of the solar image. The quartered disc so produced can be read like a clock face and positions and angles estimated from the centre, either horizontally or vertically from the cross-lines. Sunspots can then be transferred to the drawing paper, with its circle divided into the four quadrants.

Observing by Projection

The safest, probably the easiest, and certainly the most accurate method of solar observing is by projection of the Sun's image through a telescope on to a flat white card. The procedure (shown in Figures 1.1 and 1.2, for refracting and reflecting telescopes) is simple, and is widely used by skilled observers. It is also the method used by those people who want to determine accurate positions of features on the Sun's disc.

I should probably mention at this stage that it is when observing by projection that the refracting telescope becomes the ideal instrument for solar observing. It is possible to dispense with the aperture filter, so vital when viewing the Sun by the direct method with all types of telescopes.

For projection work, refractors with focal ratios of *f*/12 and over are the best. The aperture of the object lens should not be stopped down, as maximum resolution is just as important in solar work as in any other branch of observational astronomy (and the limit of resolution depends on the aperture of the objective lens or primary mirror). Small, good quality refractors of 50 mm to 76 mm are perfectly adequate for the majority of solar observations; 80 mm to 130 mm are even better, and the latter can resolve 1" of arc (into which a number of solar features fall) on days of very good seeing.

Some thought should also be given to the type of eyepiece used for projection work. Huyghenians or Ramsdens are very suitable, as they contain no cemented elements. In more sophisticated eyepieces it is possible for the concentration of the Sun's heat to melt the balsam holding the lens elements together.

Figure 1.1
Observing the Sun's image by projection, with a refractor.

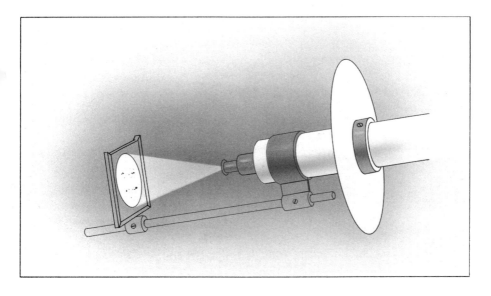

When projecting an image of the Sun on to a card, extraneous light must be excluded as much as possible from the surface of the projection card, which is of necessity mounted some distance behind the eyepiece.

This can be done in three ways: (i) attach a cardboard or hardboard baffle fitted over the telescope tube; (ii) make a projection box to fit over the eyepiece tube; (iii) use a darkened observatory. Figure 1.3 shows two types of projection box.

When using a projection box the dimensions should be sufficient to allow for a 152-mm (six-inch) diameter solar image to be projected on to its rear surface. This also applies to the setups shown in Figures 1.1 and 1.2. The Sun's image should be projected on to a surface which is mounted exactly perpendicular to the axis of the telescope, and the surface should be somewhat larger than the disc image so that the entire Sun can be displayed. Some provision has to be made for the distance between the eyepiece and the projection surface to be adjusted. We need an image that is exactly six inches in diameter, and the Sun's angular diameter varies slightly

Figure 1.2
Observing the Sun's image by projection, with a reflector.

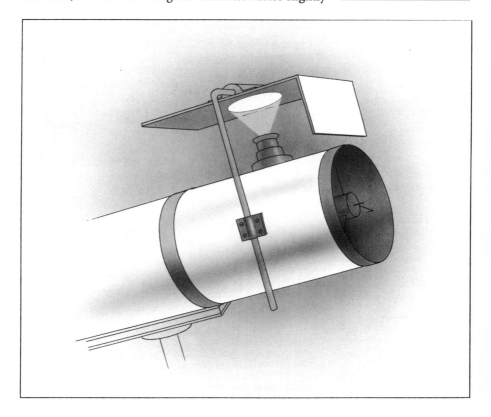

as the Sun-to-Earth distance changes throughout the year. The correct distance is easily found by experiment.

Looking at a six-inch solar image, it is immediately apparent that the disc is not equally bright all over, but that it shades off darker towards the limb. This is known as 'limb darkening' and is due to the Sun being gaseous, so that in the centre of the disc we are looking further into the hotter, and therefore brighter, interior. Sunspots, if present, can readily be seen, as well as irregular bright areas or 'faculae'. With a high power eyepiece, 'granulation' over the whole of the Sun's surface may be seen. These phenomena are shown in Figures 1.4, 1.5 and 1.6.

The main task at this stage of observing will be to observe and count the number of sunspots which occur daily on the Sun: and indeed if this is to be the limit of your interest then it is still valuable. Sunspots are cooler, relatively dark areas on the Sun's bright surface (the

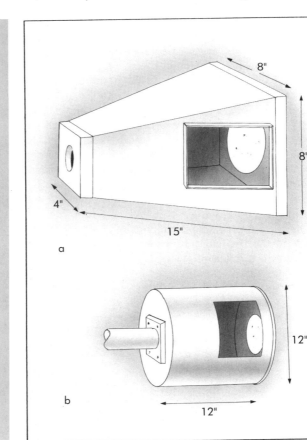

Figure 1.3 Two types of projection box: **a** is made of balsa wood; **b** is a metal drum.

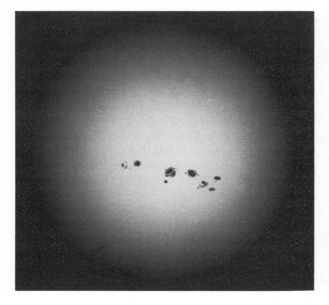

Figure 1.4 The Sun's disc, showing limb darkening and large groups of sunspots.

29 April 1984; 5-inch refractor

photosphere), and are formed where magnetic fields inside the Sun break through the surface and loop up into space. The frequency of sunspots and other forms of solar activity rises and falls in the (approximately) 11-year solar cycle, and by counting the number of spots observed daily, and at the end of each month dividing the number seen by the number of observing days, the Mean Daily Frequency (MDF) is thus derived. This can be plotted on a graph to show the rise and fall of solar activity obtained from sunspot counts throughout the cycle. Figure 1.7 a and b shows activity for Cycle 22, the present cycle.

There are two principal methods of making sunspot counts in the UK. Observers can use either the count based on the number of Active Areas (AAs) seen on the disc, or the internationally used method known as the Relative Sunspot Number.

Active Areas (AAs): Every spot, however small, counts as a separate Active Area if it is at least 10° of latitude or longitude from its nearest neighbour. The same rule applies to spot groups.

A large group, however spread out in area, is still one active area unless it has distinct separate centres of activity at least 10° apart.

However, from time to time distinct groups do break out nearer than 10° from each other. When such groups

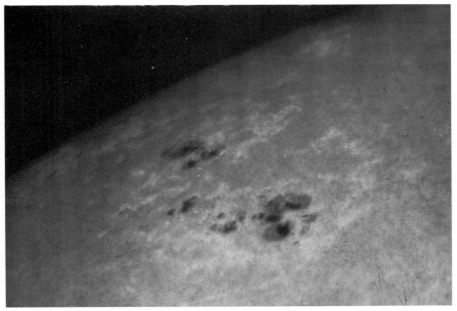

Figure 1.5
Faculae at the Sun's limb, with groups of sunspots embedded in them.

18 February 1969
1056 UT; 5-inch OG
×210 $^1/_{1200}$ s

occur they should be counted as two AAs and a note made that this has been done.

Relative Sunspot Number (R): To find the Relative Sunspot Number, proceed as above, then count all the spots that you can see either singly or in groups.

Assume that the Relative Sunspot Number = $10g+f$, where g represents the Active Areas you have counted (AAs), and f the number of spots counted within the AAs. So the calculation is simply to multiply g by 10 and add f to get R for each day. At the end of each observing month calculate the monthly R for your observations by dividing the total of your R column by the number of observing days. One of the major tasks for solar astronomers, amateur or professional, is to plot the positions of all the visible sunspots as they vary from day to day over the whole sunspot cycle. The six-inch (152-mm) diameter disc has now become the standard for British Astronomical Association (BAA) observers. The projection card should therefore have a six-inch diameter circle drawn on to it, divided by fine pencil lines into $^1/_2$-inch (12.5-mm) squares and diagonals. These lines should be drawn faintly so as to not hide any small sunspots. The squares must be readily identifiable, so they are numbered, down the left side from 1–12 and across the top (from left to right) A–L. Figure 1.8 shows a prepared grid.

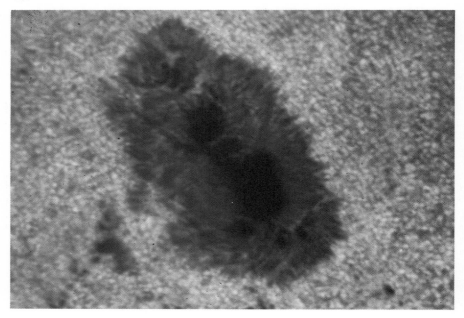

Figure 1.6
Surface granulation, and a sunspot.

11 June 1969 0714 UT; 13-cm refractor ×210

Make an identical grid with heavier pencilled lines on strong white card, and mount this on a small hand-held drawing board. Prepare a number of blanks on thin (semi-transparent) paper with six-inch circles, marked at the centre and the North, South, East and West cardinal points.

You can then lay one of these sheets over the grid (which will show through), and copy any spots seen on the faint projection card at the telescope on to it.

Begin observing by fitting a suitable eyepiece and adjusting the projection card distance from the eyepiece, to ensure that the Sun's projected image exactly fits the drawn six-inch circle with its grid lines. With most telescopes the projected image has N at the top, S at the bottom, E to the right and W to the left. Some telescopes (for example, Newtonian–Cassegrains) will differ from this.

With the telescope drive turned off the image on the card will drift from right to left. If a sunspot is present it should be brought to one of the E–W lines by using the declination slow motions, then moved backwards and forwards by the RA motion. By rotating the whole projection arrangement, the card can be correctly oriented E–W, so that the spot will accurately follow the E–W line.

If a spot is not visible, then the N or S limb can be used for this adjustment, although less accurately. With an altazimuth mount (not recommended!) this proce-

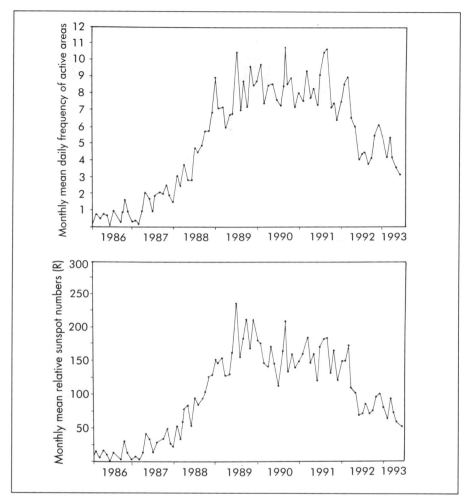

Figure 1.7 Solar activity graphs showing: **a** monthly mean daily frequency of AAs; **b** monthly mean R.

dure has to be repeated every five minutes to ensure sufficiently good orientation. An equatorial mount should ensure that the image remains correctly oriented throughout the time of the observation, but it is still useful to make frequent checks on the orientation, as described above.

Once having achieved the correct alignment and orientation of the solar image within the six-inch diameter circle on the projection card, you can copy the positions of any sunspots which are visible on the faint grid on to the thin paper overlaying the heavy lined grid mounted on the drawing board. Don't forget to make a note of the time (UT – Universal Time) and of the seeing conditions of the observations.

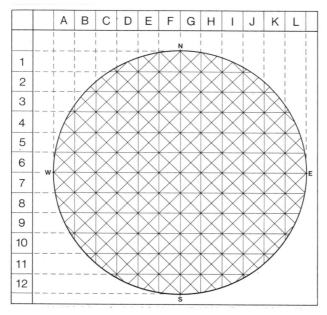

Figure 1.8 The locations of sunspots projected on to this grid can be marked, and transferred to a standard six-inch-diameter blank disc.

An alternative method uses a sheet of graph paper divided into 2, 10 and 20-mm squares. The position and size of the spots are recorded directly on to the paper at the telescope. After the observation has been made the spots are copied on to another sheet of graph paper to keep, the original drawing made at the telescope is erased, and the graph paper is used again for recording the next observation. One sheet can last for up to twenty days.

Whichever method is chosen to record positions, the next step is to fit a higher power eyepiece and slowly scan the whole of the Sun's enlarged disc (projected now on to a clean white card). It is often possible to detect small pores which perhaps were originally missed but which, once seen, can be found on the six-inch circle and recorded. Very faint spots and pores can sometimes be identified by gentle side-to-side movements of the white card, since the eye is more sensitive to a moving object. This technique also helps to eliminate any irregularities in the card being take for faint spots. Some observers just tap the card gently to obtain the same result. Spotless days should, of course, be recorded as such.

A note should also be made of any striking faculae when seeing is good, and particularly of any small bright patches near the poles. Polar faculae appear as bright points of light, a few seconds of arc in diameter, on the polar caps above 60° latitude. They are not grouped so as to form luminous patches like the faculae at lower latitudes, but are scattered at random. They are best observed in a darkened observatory by projection, using a similar disc, this time 10 inches or more in diameter. The positions and shapes of faculae can be shown on the drawing by dotted lines or, better still, the facular patches can be indicated with yellow crayon.

The last stage is to fill in the remaining details on the completed drawing:

Instrument: This can already be on the prepared blank, as the same instrument (telescope) should always be used if true comparisons are to be made.

Rotation Number: The dates of commencement of the synodic rotations, in continuation of Carrington's Greenwich Photo-Heliographic series. (The rotation numbers and the date and time of commencement of each are given in the British Astronomical Association's *Handbook* and the Astronomical Almanac.)

Date: This should be given with the year first, followed by the month and then the day.

Time: Universal Time (UT) should always be used.

Coordinates: These are the Sun's coordinates for the time the observation was made, all measured in degrees.

- P is the position of the N end of the axis of rotation, with + if E and – if W.
- B_0 is the latitude of the centre of the Sun's disc, + or –, indicating the tilt of the axis towards or away from the observer.
- L_0 is the longitude of the centre of the disc or of the central meridian (CM).

Tables giving the values of all three coordinates are given in the BAA *Handbook* and the Astronomical Almanac. The value of L_0 decreases with time by 13.2° per day, and must be adjusted for the time of observation from the values shown in the tables.

A typical recording is shown in Figure 1.9.

It is sometimes necessary to define the position of a spot or other centre of activity on the Sun's disc by latitude and longitude. What are known as 'Stonyhurst discs' (see Figure 1.10) are very convenient for this purpose, being a series of eight transparent prints of the solar disc with the lines of latitude and longitude marked. Different discs are used for different dates, varying with the tilt of the Sun's axis from +7° to –7° (towards and away from the observer). These discs or grids are six inches (152 mm) in diameter and can be placed over the drawings and the coordinates read off. The true longitude is quickly calculated from the longitude of the central meridian of the day (L_o). As has already been stated, the value of L_o decreases by 13.2° per day. Because the Sun rotates from East to West, the value of L_o increases East and decreases to the West. If a spot's position is to the West of the CM by x degrees,

Figure 1.9 A typical recording of sunspot observation.

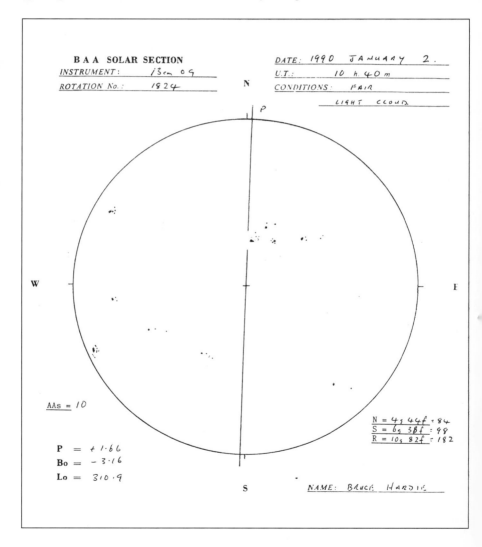

BAA SOLAR SECTION

INSTRUMENT: *13 cm 0 G*

ROTATION No.: *1824*

DATE: *1990 JANUARY 2.*

U.T.: *10 h. 40 m*

CONDITIONS: *FAIR*

LIGHT CLOUD.

N

P

W

E

AAs = *10*

N = *4, 44f = 84*
S = *6, 38f = 99*
R = *10, 82f = 182*

P = *+ 1·66*
Bo = *– 3·16*
Lo = *310·9*

S

NAME: *BRUCE HARDIE*

then that amount has to be added to the longitude of the CM to get the true longitude position of the spot. If East of the CM by x degrees, the amount must be subtracted from the longitude of the CM.

Seeing

Figure 1.10 A Stonyhurst disc.

One of the more serious limitations to solar observation, which has its origin in the Earth's atmosphere, is 'seeing'. Seeing is the quality of image, which is subject to degradation by fluctuations of the refractive index in

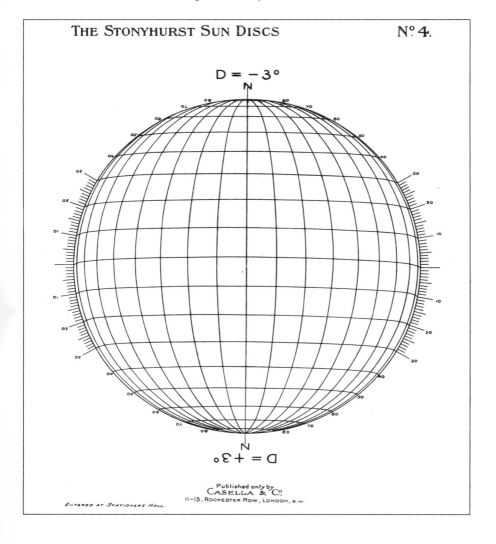

THE STONYHURST SUN DISCS　　　　　　　　N°. 4.

D = −3°

N

D = +3°

N

Published only by
CASELLA & C°.
11–15, ROCHESTER ROW, LONDON, S.W.

ENTERED AT STATIONERS HALL.

the light path through the atmosphere. As such, of course, it is a general astronomical problem.

However, in the context of solar observation the difficulties are aggravated by the nearness of the Sun itself. The Sun causes thermal convection in the entire troposphere; it heats the ground around the observatory and thus produces local convection, it heats the observatory building, and finally it heats the telescope into which it must shine. Any heated surface is a source of air instability. The viscosity of air is low, so the resulting motion caused by convection will be turbulent. The concomitant temperature fluctuation, up to 0.1° in free atmosphere but frequently much larger in and around the building and the telescope, affects the refractive index and hence generates wave-front aberrations.

Image degradation resulting from poor seeing is a very complicated process, but often three different aspects can be identified. *Blurring* is the defocusing effect of air having an index of refraction which differs from place to place. The whole image loses its sharpness. If the image essentially remains sharp but is rapidly shifted back and forth, we speak of image *motion*. If substantial parts of the image remain sharp but are shifted relative to each other, then we have image *distortion*.

Not much can be done to improve seeing conditions; as an important part of the problem originates in the free atmosphere, it is outside the observer's control. Many observers find that the best seeing is obtained in the early morning before the Sun has heated up the ground. When seeing is really bad, stopping down the aperture of the telescope can often improve matters.

The BAA Solar Section grades seeing on a scale of 1 (excellent) through to 5 (bad). This is of course somewhat arbitrary, but the following criteria correspond to the scale:

1. Clear fine structure in penumbra, granulation very well visible, no limb motion visible.

2. Some fine structure in penumbra, small spots within groups easily seen, granulation visible, slight image motions at limb.

3. Granulation barely visible, umbra–penumbra limit clear, limb motion boiling.

4. Umbra and penumbra of large spots separable, no granulation visible. Some bad image and limb boil-

ing. Grade 4 is about the limit for useful work; heavy, drifting cloud also influences this category.

5. Large spots only can be seen, bad boiling, cloud; umbra and penumbra cannot be distinguished.

Another, and perhaps more objective, measure of sunspot activity is provided by the area of the spots; however, even the measurement of this quantity is subject to some uncertainty. The relation between area and sunspot number for a single day is rather loose, but the monthly averages reveal a close connection which takes the form:

$$A = 16.7R$$

where A is the area in millionths of the visible hemisphere, corrected for foreshortening. Thus, near the peak of the solar cycle, when the value of R might be 100, the total area of all spots on the visible disc would be 1670 millionths.

A convenient way of measuring the area of sunspots uses 1-mm graph squares for a solar diameter of 600 mm. The spots can be projected or drawn on to the graph paper. Tracing paper can also be used, or a transparent graticule on plastic can be prepared to lay over a drawing or photograph.

The number of small squares covered by the spot, or group, are counted and multiplied by 1.72 to give the area in millionths of the Sun's visible hemisphere. A correction has to be made for foreshortening, unless the group is at the centre of the disc, and this is conveniently calculated from the position of the group on the 6-inch disc drawings, the distances of the spot from the centre of the disc being measured in inches.

Classification of Sunspots

Sunspot groups can be classified according to their evolutionary development. We begin by looking at various forms of sunspots and sunspot groups in terms of the 'Zürich classification' (Waldmeier M, *Ergebnisse und Probleme der Sonnenforschung*, 2nd edn, Geest u. Portig, 1955). This classification is largely a time sequence which, if all stages of it come about, may take several months. It distinguishes unipolar and bipolar configurations, the

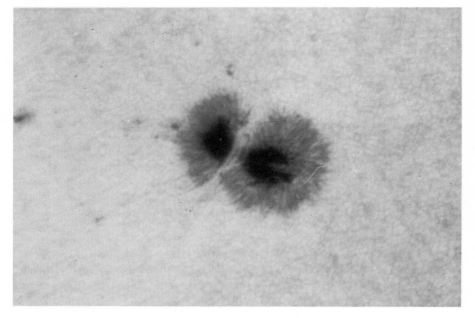

size and complexity of the spot/group, and the presence or absence of a penumbra, i.e. the ray-structured halo of intermediate intensity which in general surrounds the umbra, of large spots (see Figure 1.11).

A. A single spot, or a group of spots appears. There is no penumbra, and no bipolar configuration.

B. A group of spots without penumbra. The group is dominated by two spots that mark the two magnetic polarities, and the bipolar character is clearly noticeable.

Figure 1.11
Delicate detail in the penumbra of a large sunspot.

8 July 1983 0944 UT; 5-inch OG, $1/_{500}$ s

The direction of bipolar configurations is roughly East–West on the Sun, but with the leader spot (in sense of rotation) slightly closer to the equator than the following spot (Joy's law). The inclination of the pf axis (for 'preceding' and 'following', respectively) decreases during the early part of the group's evolution. This is largely as a consequence of the group's growth in the East–West direction.

For more than half of the sunspot groups, the evolution terminates, after a day or a few days, with state A or B. They never develop penumbra. For those that do, the Zürich classification continues:

C. Bipolar group; one of the principal spots has penumbra.

D. Bipolar group; both principal spots have penumbra, but at least one of them still has simple structure. The group extends over less than 10° on the Sun.

Mostly, but not always, the larger spots have penumbra, while the smaller ones have none. Spots have been seen without penumbra as large as 1100 km in diameter, and spots with penumbra as small as 2500 km in (umbral) diameter. The typical radial extent of the penumbra is 5000 km. The penumbra forms out of radially aligned granules surrounding the umbra, a process which may be completed within one hour. Again, growth of a sunspot group may end with state C or D. On the other hand there are further growing groups, classified as E and F.

E. Large bipolar group, extending over more than 10° on the Sun. Both principal spots have a penumbra and often a complex structure. Numerous small spots are present.

F. Very large, and very complex, bipolar group, extending over more than 15°.

The following three classes describe the decay of the group, and of the last remaining spot.

G. Large bipolar group, of extent >10°. The small spots in between the principal ones have disappeared.

H. Unipolar spot with penumbra, diameter (including penumbra) >2.5°.

J. As H, but diameter <2.5°.

The H spot is often rather symmetrical, and stays without change for several weeks, except for a slow decrease in size. It is known as the 'theoreticians' spot'. It has been demonstrated that the slow decay of the spot's area, A, proceeds at a rate which is nearly constant, and is the same for a large number of studied cases. Waldemeier (1955) has provided the following rule-of-thumb for determining the approximate lifetime of a spot group in days, T, according to its area:

$$T = 0.1 A_{max}$$

where T = lifetime in days and A_{max} = maximum area in millionths of the visible hemisphere. Thus a group which attained a maximum area of 400 millionths would have a lifetime of 40 days.

McIntosh (1981) has added two more parameters to the Zürich classification. These parameters have to do with the shape and complexity of the largest spot of the group, and whether it is 'compact' or 'open', i.e. densely or sparsely filled with spots. The additional classification has proved valuable in the prediction of flares: the more complex the configuration, the more likely is the magnetic instability which gives rise to flares.

The McIntosh classification is based on a 'Modified Zürich Classification', which combines the original category G with E and F, and J with H (see Figure 1.12).

To this modified classification McIntosh attached the following sub-classifications for sunspot groups, expressed by the second and third letters of a three-letter code, the first letter being the Modified Zürich category (see Figure 1.13). The second letter indicates the type of penumbra which surrounds the largest spot of a group.

x. No penumbra. The width of the grey area bordering

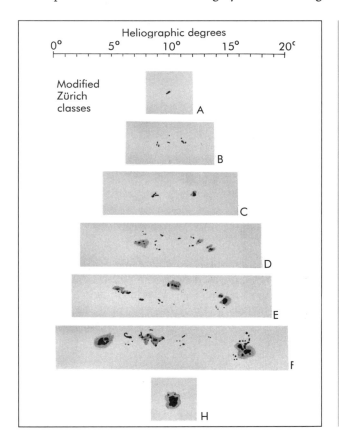

Figure 1.12 The 'Modified Zürich Classification', which classifies sunspot groups according to their evolutionary development.

spots must exceed three arc seconds in width in order to classify as penumbra.

r. A simple, or rudimentary, penumbra; the penumbra is incomplete or irregular. Rudimentary penumbra represents the transition between photospheric granulation and filamentary penumbra. Recognition of rudimentary penumbra will ordinarily require photographs or direct observation at the telescope.

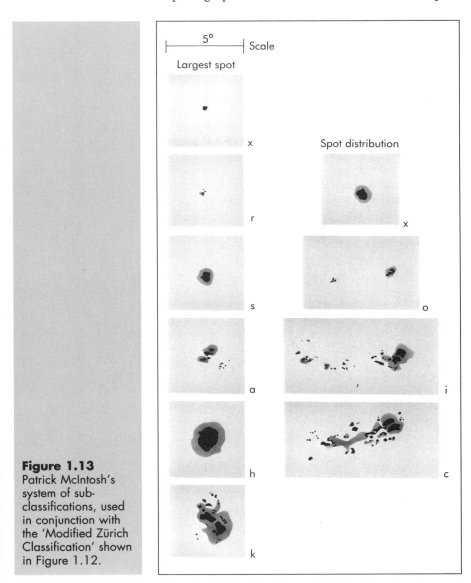

Figure 1.13
Patrick McIntosh's system of sub-classifications, used in conjunction with the 'Modified Zürich Classification' shown in Figure 1.12.

s. Symmetric, nearly circular, penumbra with filamentary structure and a spot diameter not exceeding 2.5 heliocentric degrees. The umbræ form a compact cluster near the centre of the penumbra.

a. Asymmetric, or complex, penumbra with filamentary fine structures and a spot diameter along a solar meridian not exceeding 2.5 heliocentric degrees diameter. Asymmetric penumbra is irregular in outline. or very elongated, with two or more umbræ scattered around it.

h. A large, symmetric penumbra with a diameter greater than 2.5 heliocentric degrees.

k. A large, asymmetric penumbra with a diameter greater than 2.5 heliographic degrees. Other than size, its characteristics are the same as category 'a' above. When the longitudinal extent of the penumbra exceeds 5 heliographic degrees, it is almost certain that both magnetic polarities are present within the penumbra, and the classification of the group then becomes Dkc, Ekc or Fkc.

The third letter describes the distribution of individual spots within the sunspot group:

x. Single spot.

o. Open spot distribution; the area between leading and following ends of the group is free of spots so that the group appears to divide into two areas of opposite polarity. An open distribution implies a relatively low magnetic field gradient across the line of polarity reversal.

i. Intermediate spot distribution; some spots lie between the leading and following ends of the group, but none of them possess penumbra.

c. Compact spot distribution; the area between the leading and following ends of the spot group is populated with many strong spots, with at least one interior spot possessing a penumbra. The extreme case of compact distribution has the entire spot group enveloped in one continuous penumbral area. A compact spot distribution implies a relatively steep magnetic field gradient across the line of polarity reversal.

Solar Flares

Patrick McIntosh's classification system, shown in Figure 1.13, is most useful for flare observations, because it indicates which sunspot groups are most likely to produce a flare. Based on flare observations over the last twenty years or so, groups of the modified Zürich class F have a 60% chance of producing a flare within the next twenty-four hours. Groups that have a largest spot of category k have a 50% chance of flaring; groups with a category c spot distribution increase their chances of flaring to 70%. Therefore the highest probability of seeing a flare occurs for groups of type Fkc, Fki and Fsi. When these categories are present on the Sun's disc it might be a good time to watch out for the rare, white-light flares, but chances are small unless some time is spent monitoring the spot group.

Flares are short-lived phenomena which can be easily missed. A small video camera, attached to the end of the telescope and coupled to a video recorder, is worth trying. This author has used a video camera monitoring an enlarged, projected image of a spot group, in close-up on a white card in a darkened observatory, although so far no flares have been seen with this setup. It should be noted that only white-light flares may be detected by these means; the main observation of flares is carried out in Hydrogen-alpha light.

References

Baxter W M, *The Sun and the Amateur Astronomer*, Cambridge: Cambridge University Press (revised edition 1973)

Giovanelli B, *Secrets of the Sun*, Cambridge: Cambridge University Press (1984)

Mitton S, *Daytime Star*, Cambridge: Cambridge University Press (1981)

Noyes R, *The Sun: Our Star*, Harvard (1982)

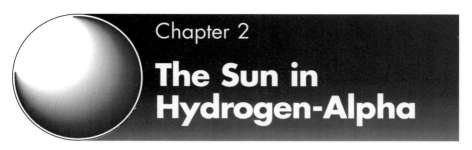

Chapter 2

The Sun in Hydrogen-Alpha

Bob Turner

Until the turn of the century, the only times when the inner solar atmosphere could be observed were the fleeting moments of totality during eclipses of the Sun, when the whole of the solar corona and the prominences on the Sun's limb are viewable. Since the advent of the spectrohelioscope and sub-Ångström filters, each observation can now become a total eclipse of its own.

The most common viewing frequency for amateur astronomers is at 656.3 nm, the hydrogen-alpha line, although filters are available to allow the Sun to be viewed in hydrogen-beta at 486 nm, or in other lights such as calcium.

The view of the Sun with a 4-Ångström bandwidth filter will show limb prominences, but it requires far more expensive filters of shorter frequency to view solar surface activity.

The use of a sub-Ångström (bandwidth) filter, although very rewarding, is not something that can be accomplished immediately. The delicacy of the instrumentation means that a considerable period has to be spent in actually learning how to observe. Most of the filters on the market allow their fine tuning to be carried out by slightly tilting the filter, and therefore a very delicate touch is necessary to obtain the best possible view.

The H-alpha line is not at a single frequency, but has a bandwidth that can be scanned, and certain objects show up better in some parts of the band than others. Scanning from the 'red wing' to the 'blue wing' will show a marked difference, as the rotation of the Sun will give a shift in the line. Prominences tend to show up better just before the peak of the H-alpha line, while surface features, filaments and granulation tend to be slightly better after and during the sweep of the peak across the filter.

The filters work on a polarising series of elements, and therefore 'seeing' can be improved by rotating the filter through 90° to find the best possible observation point.

Such filters are heat sensitive and will slightly change frequency when warmed by heat from the solar image. Also the optical configuration itself can change focus, and therefore sub-Ångström viewing is a matter of continual readjustment to obtain the best view for the phenomenon under observation.

Observing Solar Activity

Every time you observe the Sun there is bound to be some prominence activity on the limb that can be recorded. The prominences can be of several varied forms. The longest-lasting prominence type is the 'hedgerow', which seems to hang in the solar atmosphere without appreciable change for a considerable length of time and can be seen over periods of up to several rotations. These prominences are mainly diffuse and delicate in structure, and appear dark red in colour, superimposed on the black background of the sky. When any prominence is viewed against the solar disc it is termed a filament and a very rewarding observation is to watch, over a period of several days, a 'hedgerow' prominence rotate on to or off the disc. Sometimes, if these prominences are cross-lateral, they can be viewed as part-prominence and part-filament, which gives a wonderful three-dimensional effect.

It is very difficult to interpret exactly what is being viewed at the edge of the disc, as prominences could originate on or behind the limb, or a mixture of both. 'Hedgerow' prominences are often seen as a series of arcs showing the magnetic structure of the Sun in the lower corona. When these loop-type bridges are foreshortened, it can be difficult to interpret exactly what it is you are looking at.

Some prominences, known as spray prominences, are combined with highly active areas and will spray material at a velocity of up to 2000 km/sec. This exceeds the escape velocity of the Sun, and thus the structure of this phenomenon can change rapidly over the course of a few minutes. Watching spray prominences is rather like watching the minute hand of a watch: it is not actually possible to see movement, but you are aware after a very short period that a change has taken place.

The more active the prominence, the brighter red the event appears in H-alpha, and sometimes the most active areas – small pinpoints of matter – can take on an orange or yellow hue depending upon the level of activity. When these active regions are associated with flares, it is possible to see the flare itself rise up the prominence's column. Although this activity is usually very short-lived, it can produce a remarkable sight.

Filaments on the disc, when seen as surges, appear black and can take on a feathery or wispy look, depending upon the amount of material being ejected and the angle at which the ejection is viewed.

The details of sunspots look somewhat different from the white light view when seen in H-alpha. The penumbra of a spot is not nearly so prominent as it is in white light, and the magnetic structure of darker or black lines shows more readily around spots or groups of spots.

Solar flares are usually associated with active spots, especially after the spot has passed its maximum growth and is in the early stages of decay.

Flares on the solar disc in H-alpha appear to be yellow or yellow/white and can be very distinctive, modifying their appearance over a matter of minutes. A spot or group of spots can spawn dozens of flares, some lasting for up to several hours, and changing all the time. Other areas will produce flare after flare in the same area, the flares only lasting a few minutes with quiescent periods between them.

Solar flares tend to happen at the magnetic interchange between the spots and will often form ribbons on each side of the darker inversion line. It is quite common to see flares appearing rather like strings of pearls along a flare ribbon, with the flare 'beads' themselves moving within the ribbon.

One of the most dramatic observations is a flare on or near to the solar limb as the obscuration of material can produce some very unusual effects, with the flare changing in brightness and position within the limb material.

Current research on flares seems to show that there is a periodicity of between 152 and 154 days, which coincides with the periodicity of soft X-ray peak flux and hard X-ray emissions. This solar cycle also shows sub-harmonic cycles of approximately 72 days and 55 days. The cause of this periodicity is at the moment unknown, but the level of prominence material on the solar limb may also show a similar periodicity.

Photography

Photography is very difficult in H-alpha. Film has to be obtained which is sensitive to the far end of the spectrum. Also, while disc details can be captured with exposures of 1/125th or 1/250th of a second, to obtain a very good photography of prominences the disc must be heavily over-exposed; intervals of $1/4$ to $1/2$ second may be needed to obtain adequate results.

As solar observers, not only can we look at the Sun, marvelling at its features and enjoying the beauty of eclipses, we also have the opportunity to look closely and at first hand at the workings of the physics of Earth's nearest star.

References

See Chapter 1, p21.

Chapter 3

Eclipses

Michael Maunder

It is a mistake to regard eclipses as phenomena which solely affect the Sun and the Moon. The passing of any one celestial body in front of another is an eclipse, in its literal interpretation. In between the rare solar and lunar events, do not forget that lunar and planetary occultations, transits, and some variable star observations, fall into the same class, and are equally important observations. Details of these will be found in other chapters of this book.

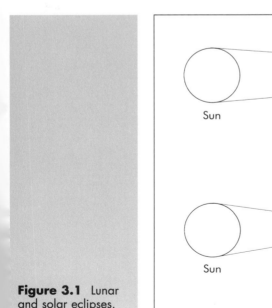

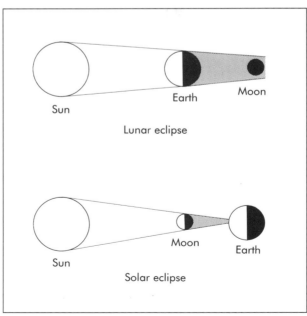

Sun

Earth   Moon

Lunar eclipse

Sun

Moon   Earth

Solar eclipse

Figure 3.1 Lunar and solar eclipses.

Geometry

To the layman, eclipses tend only to be associated with the Sun or the Moon. They occur whenever the Sun, the Moon and the Earth are in line with each other, as shown in Figure 3.1.

Although they are totally different to each other in character, the apparent size in the sky of the Sun and Moon, to us on Earth, is almost identical, at about half an inch in diameter. This means that when they are lined up with one another in an eclipse, either the Sun or the Moon can appear to vanish from the sky. Eclipses were the cause of consternation in history, but why the (so-called) enlightened modern public attributes such significance to mere alignments of heavenly bodies remains a mystery. Perhaps it is a hang-over from the calendar significance of the (unseen) near misses we call new and full Moon.

These near misses are by far the more common events, because the orbits of the Moon around the Earth, and the Earth around the Sun, are generally not quite in the same plane. For a true eclipse to happen, those two orbital planes must be the same.

Strictly speaking, the determining factor is the Moon's orbit. Where this intersects the ecliptic, as the Moon is travelling North we have the ascending node, and conversely the descending node when the Moon is travelling South. To get an eclipse of the Moon, it has to be very close to one of these nodes at full; a solar eclipse occurs with the Moon at new. For these coincidences to result in totality is very rare. Partial eclipses are not very common, either.

Because of the cyclic nature of the coincidence of the two orbital planes, and the calendar of new and full Moon, it is to be expected that a pattern will emerge. This is now known as the 'Saros'. Depending on the number of leap years it contains, the Saros is 18 years + 10 or 11 days + about a third of a day. Several Saros cycles run at the same time, and each can continue for several hundred years. Eclipse study has gone on for a very long time – it is a sobering thought to realise that Stonehenge was being built and rebuilt for a thousand years.

There is an obvious major difference between lunar and solar eclipses: the time of day at which they may be observed. A complete removal of the sunlight is unlikely to be overlooked, whereas lunar eclipses tend to occur

Figure 3.2
Progress of an
annular solar eclipse,
photographed at
10-minute intervals.

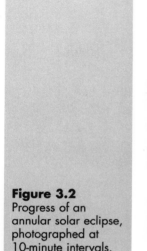

when most of the population is asleep. Lunar eclipses also vary considerably in the degree of dimming, and a total might not always be obvious.

Another difference between the two phenomena is in the ease of seeing them from any place on Earth. A lunar event can be seen from anywhere when the Moon is above the horizon, and can be total for an hour, with the partial phases lasting several hours. The different geometry for a solar eclipse means that the shadow is sharply focused, and moves very fast. Any location can expect, at best, to see a total eclipse of the Sun once every 300 years or so, and that can never last for much longer than seven minutes. Sometimes the shadow focus lies above the Earth's surface, which means that only the central part of the Sun's disc is obscured, producing an annular eclipse. Figure 3.2 shows the progress of an annular eclipse photographed at ten-minute intervals. Partial solar eclipses are only visible from a fraction of the daylit Earth.

Lunar Eclipses

Quite why these events should be considered second best is odd. Perhaps it is because they can be seen from your own backyard, with no need to visit an exotic locale. Nevertheless, the skill needed to record one on film is no less taxing than for the solar variety, and even more luck is necessary with the weather because of the lunar eclipse's fainter image.

Three distinct phases can be seen during a total lunar eclipse. The first phase is often undetectable, as the Moon enters the Sun's penumbral shadow. It is not until this shading reaches the halfway point that most observers comment on it. By the time it has fully covered the disc, the Moon is significantly darker. As the darker umbral shadow encroaches, the normal moonlight is gradually blotted out during the second phase. This is very much like the monthly lunar phases speeded up, until only a very thin crescent remains. As this process takes about an hour there is plenty of time to prepare for the third, or total, phase. After about an hour of totality, the whole sequence unwinds again, until the Moon is revealed once more in its normal full brilliance.

No two lunar eclipses are the same. The weather is a determining factor, on two counts. Local weather is pretty obvious, since it determines the sky's clarity; more important is the sky clarity all around the Earth's sunlit horizon. Clouds and moisture redden sunlight, very much as in a normal sunset, and this reddens the Moon's disc. Too many clouds, and a grey effect prevails, even to the extent of it being called a 'blue moon'. Sometimes the Earth's atmosphere has a lot of dust in it, which can darken the shadow even more. The recent eruption of the Pinatubo volcano was a classic of this; the Moon appeared to vanish completely, only being seen with any confidence through binoculars, and then only if you knew where to look for it!

Science

There is much science to be done during the event, but the novice is strongly advised simply to enjoy it all the first time round. Because we are talking about a longish period, usually in cold weather and at night, it is worth considering turning an observation session into a party. Invite the neighbours in. That way you stand less risk of security lights and other hazards to observation being left on.

As most events are not total and only penumbral, light pollution is a major hazard. Concentrate on recording timings when shadows appear real rather than predicted, and by how much the lunar surface is covered. If good magnification is available, note not only the timings but record how familiar features differ. This work is rarely attempted, and a novel crater shading observation is well worthwhile and interesting.

If there is an umbral phase to the eclipse, more serious work can be done. The whole range of timings should be attempted, from craters entering shadow through duration to exit, to the major events such as the start and end of the shadow on the disc itself. Full observational programmes are readily available from the national societies.

Perhaps the major work, and the biggest problem, is assessing the depth of darkening and its colour. Even video technology has to compress the evidence somewhat, but that technology is the nearest we have to an absolute measure. Otherwise we need to fall back on the L measures of the Danjon Scale, running from L=0 to L=4. A typical dark L=0 was the 'Pinatubo' eclipse of 1992, when the Moon was almost invisible at mid-eclipse. At the other extreme, L=4, the Moon remains extremely bright, exhibiting a coppery-red to orange colour with a characteristic bluish and very bright rim. Most people would not notice this as an eclipse.

A good experiment to carry out during totality is a magnitude estimate. The light difference is huge, dropping from around –12.5 by at least eight magnitudes. In 1992 the drop was about 17 or 18 magnitudes, as was easy to estimate from the large number of fifth magnitude stars nearby.

Do not forget that totality is an excellent time to carry out other serious astronomy, such as deep sky photography and observation. Also, lunar occultations are considerably easier with the wider range of fainter stars which become visible. Some workers concentrate on this serious side, as it allows observations to be carried out at a time when they would normally be blotted out.

Still and video photography of the lunar eclipse demand the same attention to detail as for solar eclipses, and the two are considered later in this chapter.

Solar Eclipses

The first point to make clear to anyone intending to view a solar eclipse, total or not, is the need for safety. Sunlight is dangerous and can easily damage our eyes.

This warning cannot be repeated often enough: *no filter is entirely safe when placed behind the main mirror or lens and before the eye. Never consider using such a filter, whatever the type or source.*

Filters placed in front of the main optics – telescope or camera – must have the right properties. Ultra-violet (UV) radiation will pass through a surprisingly large number of visually opaque filters. The main damage caused by UV is to the cornea, leading to cataracts and similar irreversible optical defects. Infra-red (IR) radiation is even more prone to passing through visually opaque filters. It is heat radiation, and this is concentrated on the retina or in the eyeball fluid. The retina has no pain sensors, so it can be cooked, literally, before you are aware of the damage, with irreversible vision loss in the areas affected.

A number of different materials may be encountered in solar filters. These are the safety points to watch for.

Mylar: Mylar filters have become the most popular type for solar work in recent years. Mylar is an extremely tough plastic film. The type to use is 10 microns thick, coated with a very thin layer of aluminium metal, preferably on both sides. Aluminium absorbs both UV and IR radiation and transmits only a small amount of visible light, mainly blue. Only use Mylar which has been obtained from a reputable source.

Mylar film must never be confused with 'silver paper', which is a thin sheet of aluminium foil; nor with glitters, which are similar plastic sheets but with a silver or metallic ink printed on to them. These are also becoming very common in the packing industry, and are very dangerous indeed because they allow too much damaging radiation to pass.

Mylar itself is extremely tough, but the aluminium coating is not. Each piece must be inspected for pinholes. These holes transmit a lot of damaging radiation, as in a pinhole camera, and also cause image degradation due to halation. This is seen as lowered contrast and sharpness. The surface must be inspected regularly for unevenness. Scuffing and other abrasions will give local highspots and pinholes. The dangers are obvious and image degradation can be appreciable. Some slight rucking or unevenness will not cause image degradation. On the contrary, some authorities positively recommend not stretching the film too tightly.

The main problem with Mylar arises from a factor which is rarely mentioned or discussed: the film is 'poled'. In other words, it has a marked polarisation. You can see this by holding up a piece of Mylar to sunlight. The Sun's image will often seem to have 'wings'. These rotate as the piece is turned round. The effects really become serious with some types of camera viewfinder, particularly in autofocus cameras where circular polarisation is needed. Before attempting to photograph anything important, check the focus by rotating the filter, and always double check with test exposures.

The colour image with a Mylar filter is blue, which can be unacceptable in some situations.

Silver printed plastics: Filters made of this material *should not be used under any circumstances.*

'Negative' film: 'Black' colour film *should not be used under any circumstances.* 'Black' monochrome film is safe to use if the material is dense enough. It gives better colours than does Mylar. The image is rarely easy to focus and has low contrast due to halation, arising from the Callier Effect.

Wratten No. 96 filters: These are optically the best filters in the visible range for pre-set or rangefinder cameras, i.e. where the image is not seen directly. However, these neutral density filters are totally transparent to infrared radiation and images produced through a Wratten 96 filter by direct means, such as through a normal single-lens reflex (SLR) camera's viewfinder, *must not be viewed with the eye under any circumstances.*

Inconel and other metal coatings: Several commercial versions exist of this metal coating on optical glass. They are the best available, and are produced to precise optical specifications. The drawback with them is their cost.

Partial Solar Eclipses

By far the safest way to view any solar eclipse is by projection. Almost any telescope or pair of binoculars can be used, and Figure 3.3 shows ingenuity at work.

Even simpler to use is a pinhole. 'Pinholes' occur in nature, and many an attractive picture is made with patterns thrown on the ground from sunlight filtering through the leaves of a tree.

Apart from providing an event which generates public interest, partial solar eclipses have little scientific

value, although many would argue that they are science in action. Serious visual observers usually limit the science to timings of the first and last moments of shadow contact. First contact is not as easy to record as it would seem, and much effort is needed to include sunspot timings. Some observers attempt to measure sky darkening and, if the eclipse is more than 50%, some of the more subtle effects on plants and animals. The situation is vastly different for a total eclipse.

Figure 3.3
Viewing a solar eclipse, using projection and ingenuity.

Total Solar Eclipses

The extra mental tension that exists during a total eclipse of the Sun is enormous. Nothing else really matters except those precious moments of totality. Picture, if you can, the tension of all those around you as the Sun gradually vanishes. Add to this the weird lighting. Shadows are still there but they are much more diffuse, and they take on a distinctive greyish hue. In the last few minutes before totality, the sky itself near to the Sun takes on a unique, steely blue tinge, with reddish to orange shading towards the horizon. If you have a spare camera or video, it is an experiment well worth undertaking to devote this simply to recording the darkening sky. The modern automatic exposure systems cope very well and the result will give an impression to others later, bringing back powerful memories for all those who experienced the event. They are also a sensible back-up in case the weather turns nasty during totality.

Shadow Bands

If you are really lucky, with crystal clear skies, shadow bands might occur. These are seen as a diffuse pattern of light, with darker shadings, moving rapidly over any clean flat surface. Whitewashed walls, sheets or even large pieces of paper have all been pressed into service as viewing aids. Shadow bands only appear a minute or so either side of totality. More observations are needed, with timings, to determine what they are and how they are formed. Many experienced photographers describe them as the ultimate test of skill and luck.

Baily's Beads and the Diamond Ring Effect

Both phenomena occur because the Moon has mountains and craters and is not a perfect sphere. In the few seconds before and after totality, the last vestiges of sunlight peek through the lunar valleys to create the effects.

Francis Baily, in 1836, first described the line of points of light as one with an appearance like beads, and the name has stuck. The effect is always less dramatic when seen after totality, and the name tends to refer more to the first apparition. In the event of an annular eclipse, with the Moon just fractionally smaller in apparent diameter than the Sun, Baily's Beads can encircle the disc. This is known as the Diamond Ring, or Necklace, effect. Because the full glare of the Sun is not there, a very precise measure of the Sun's apparent diameter can be made from such an observation. It has been used as a check on the real size of the Sun to see if it is variable (apparently it is not).

The Diamond Ring effect has been observed just before totality, but it is always more spectacular just afterwards. Photography confirms that impression, so it is not just an optical illusion combined with some dark adaptation of the eye. The effect is so startling, some say beautiful, that many observers travel to see just this second or two of nature at its most exquisite. The Diamond Ring effect signals the end of totality, and that full eye protection is needed again until the next total eclipse.

The Chromosphere

Just before Baily's Beads appear, eclipses display an amazing drop in light level as the Sun's limb is gradually extinguished. This is due to limb darkening on the Sun, very much like our sunsets through the thicker atmosphere of the horizon. The light gets redder, too, but that is not all the story. Just above the Sun's surface is the chromosphere, and what is seen is an almost smooth transition from the reddening limb colour to the pink of the chromosphere proper.

Seen from Earth, the thickness of the chromosphere is but a small fraction of the diameter, and in long eclipses it may be seen only at the leading and trailing edge of the Moon's disc. Some observers sacrifice maximum duration in order to capture more striking pictures, with more of the chromosphere in evidence at one of the other limbs.

When the Moon's diameter closely matches that of the Sun, the chromosphere can be seen all the way round, and is a spectacle in its own right. It is sometimes seen during annular eclipses.

Prominences

Poetically, but incorrectly, described as 'flames' or 'tongues of fire', these are the prominences and filaments studied in hydrogen alpha light by solar astronomers. It is only during a total eclipse that they can be seen with the naked eye. The colour is a strong pink to red and, being monochromatic, must be a disappointment to those of the male population with colour blindness.

The shape and structure of prominences vary enormously, as does the number seen. There is always a greater number at sunspot maximum and when spots are on the limb. Some eclipse chasers spend much time plotting sunspot groups weeks in advance, in the hope of catching prominences at the right time and place on the disc on the day. Others let nature take its course and prepare for a pleasant surprise.

The Corona

Beautiful as the other phenomena are, all observers agree that there is nothing (with the possible exception of the Diamond Ring) to compare with the corona in full display. Even hazy or cirrus clouds cannot detract from its full wonder.

No two men seem to agree on the colour (colour blindness again, and variations in blue sensitivity), but a pearly pink/grey is the most common description among women, who have better colour vision. Add to that the tracery and filamental structure spreading out at least a full solar diameter, and you get the beginning of a feel for how it appears. There is simply nothing else like it for sheer spectacle.

The overall shape varies according to the sunspot cycle, being more circular at maximum, and with sharply delineated wings at minimum with plumes and spikes at angles from the centre.

The naked eye can usually detect the corona out to more than a solar diameter (angular 1.5–2°), particularly if dark-adapted. Any form of optical aid is always recommended, and with a modest pair of binoculars the corona should be visible out to several solar diameters. It is the only way in which to appreciate fully the corona's delicate structure.

Other Phenomena

If you can tear your eyes away from the corona, spare a moment or two to look about. The Moon's shadow often appears quite clearly; the strange reddish horizon and edges to the shadow make for a scene never to be forgotten, and best seen across a clear landscape.

During, and for a good time before totality, the brighter stars and planets will appear. A wide-angle photograph is always something to attempt. Charts are always available in the major astronomy magazines to plan this activity.

Even more important to some people is the fact that during an eclipse the sky becomes dark at a time when this is not normally available. Comet hunters frequently spend this time searching the sky in the hope of repeating the discovery of 1948; others catch up on variable star work.

Photography

Even experienced photographers can make mistakes, and an eclipse simply is not like an ordinary assignment. The best advice to offer the novice is to regard photography as a bonus. Go prepared to view an eclipse with nothing more than the naked eye.

The great advantage of naked eye viewing is that there is little or no clutter, and last minute changes of plan are easy. You also have time to stand and stare. In your mind will be the everlasting picture, which no photograph ever replaces for vividness. Photographers should always come mentally prepared to abandon all their well-laid plans when the moment comes, and fall back to simple naked eye viewing. Someone, somewhere is bound to get a picture to keep as a souvenir. Snapshots of the site and the people there can also have some value later, particularly if anything unusual has happened.

Preparations for Photography

The history of eclipse trips is littered with tales of photographers who failed to get their pictures because they had not had a particular camera for long enough to get used to its funny ways. Stick with an old one if in doubt. Better still, take your old camera along as well. All professional photographers carry at least one spare.

Choice of Camera

Cost is usually the decider, and this applies to video equipment just as much as to conventional cameras. However, simplicity is much more important, and the camera's controls must be second nature. Remember, a lunar eclipse takes place at night and a total solar eclipse in what amounts to pitch darkness. Always practise under adverse lighting conditions where the controls cannot clearly be seen.

Simple still cameras, such as those fitted with a rangefinder, are possibly the best equipment to use for anyone who is only interested in filling a picture album. The simple camera's ability to cope with a wide range of lighting is legendary, but it is important to consider those observers with more serious intentions; with this in mind, do not take a camera fitted with an automatic flash. Not only will you fail to record the real lighting, you could upset a crucial picture in another camera. Worse still, you could destroy the essential dark adaptation of anyone doing serious visual work.

SLR cameras remain the most popular choice for the photographic recording of eclipses because of the wide range of lenses capable of being fitted to them. Check battery consumption, fit fresh batteries and pack plenty of spares before you leave home.

Choice of Lens or Telescope

Sheer image size is not the only criterion – quality has also to be considered. The longer the focal length chosen, the more the quality of the whole system has to be taken into account.

Do not use a zoom lens; instead, always select the best 'prime' lens you can afford. Mirror telephoto lenses are best for colour rendition, but they are usually too 'slow' for lunar eclipses. Camera shake is the main killer of image quality at eclipses. Extra-heavy lenses and attachments upset the camera's natural balance, and a very sturdy tripod becomes the most essential ingredient in the whole system.

Tripods are not cheap, at least those of the quality needed for lenses of long focal length, and money spent on a good one is often a much better investment than a fancy lens would be. Tripod effectiveness must be checked long before you leave. On the basis of this, decide on the longest focal length you should take. Whatever precautions you might take, some cameras are quite incapable of giving sharp images with long lenses because of camera shake. That is the time to consider the other end of the focal range. Many of the more interesting or artistic eclipse pictures are taken with wide-angle lenses. These take in the true scenic atmosphere and can be quite useful scientifically. The Moon's shadow will show, and some of the fascinating colours around the horizon. Many of the best pictures of this type may be taken with a 28-mm lens, but 21-mm is becoming a popular focal length. Fisheye lenses come into their own for these special events.

Choice of Film

Video films are so good that the camera's autoexposure functions will cope with virtually any situation. It is not so simple with still photography but, because of the better quality, it is worth persevering. If you have a favourite film for normal use, stick with it. The film's properties will be known to you, and will be much safer.

The general guidelines are very simple:

- Slow films give less grain and better colours. Use 100 ISO for prominences and the inner corona in total solar eclipses (see Figure 3.4), and for the partial phases of lunar eclipses.
- Fast films record faint detail where quality is of secondary importance. Use 400 ISO or faster for the corona (see Figure 3.5) or deep umbral lunar events.
- Take at least twice as many pictures as you think you will need, and then some more. Set aside all the film you think necessary for the eclipse itself, and store it separately.

Test Objects

While test exposures of the Sun are important for the partial phases, they are not very good as a final check on tripod and camera stability. Exposure times are too short, even with a decent filter.

Figure 3.4 Slow film (here, 100 ISO) is best for capturing prominences and the inner corona of a total solar eclipse.

By far the best test object is the Moon. Not only is it the right size, it also gives the right lower range of brightness for both lunar and solar events. Although its apparent motion is slightly less than that of the Sun, it is close enough to check guiding or motor drive errors.

The full Moon is about as bright as the Sun's inner corona. A few exposures will be a useful double check that the *f*/ratio of the optical system is what it claims to be. The full Moon, by definition, is a day- (and therefore Sun-) lit object. The crescent Moon is the best test object for focus and camera shake checks. Allow two stops of extra exposure to pick up crater detail, and make duplicate exposures during periods of good seeing. It will probably take several lunations to iron out all the bugs.

Do not always expect that the image in the viewfinder will be in absolutely correct focus. Make test shots a little out of focus either way, to see if the image sharpness improves. If it does, make a note on the lens barrel at the point at which actual focus occurs.

Exposure Planning

Video photography needs the same discipline as does still or movie. Plan a sequence of exposures, keep it simple, and write it down. Work through this plan in a dummy run (no film in the camera), with no more light than that of the full Moon. Allow at least 50% of the time to just stand and stare. Use that 50% now to note the

Figure 3.5 Use fast film (400 ISO or faster) for the corona.

time taken, and write it against each step. An auto-wind camera with an intervalometer makes life easier.

Keep at it until you have a plan capable of being carried out during the time of totality. Simple plans can be committed to memory, but the panic of last minute hitches makes this a dangerous policy. It is much better to take along a simple 'Walkman'-type cassette player. Use your script and a stop watch and actually record the plan, in real time, and work through the sequence using your own voice prompts. If you are well away from others on the site, another recorder to take down your verbal record of what you actually do on the day is a great idea. However, do bear in mind that other workers might not like your commentary. The advantage of your taped prompt is that it goes directly into your own ear, and blots out other voices.

Exposure Range

Knowing what to photograph at an eclipse is always the most difficult thing to decide. There are simply too many variables of equipment and weather to be dogmatic. If you can lay your hands on one of the many published lists of exposures for the different phenomena, the recommendations are usually spot on.

In practice, virtually any exposure will give a result of some sort, because of the wide latitude of modern film. It is also important to remember that the brightness of the Sun's inner corona is about the same intensity as that of the full Moon, which in turn is about as bright as an average daylit scene. Excellent pictures of the inner corona and fainter prominences are taken by setting the camera for a daylit scene and leaving the camera controls alone thereafter.

Table 3.1 gives suggested basic camera settings for total solar and lunar eclipses.

For a solar eclipse, the values in Table 3.1 will record the inner corona and prominences well enough in a

Table 3.1. Basic camera settings for eclipses

	Solar	Lunar
Exposure Index (ISO)	100	400
Focal ratio (f-stop)	16	8
Time of exposure (seconds)	1/60	1

clear sky. Lengthen the exposure by up to three *f*-stops to capture the outer corona. For a lunar eclipse, these values will capture the whole Moon in umbra. Lengthen the exposure by three stops for mid-totality, and by one stop more for each L value of darkening on the Danjon Scale, remembering that this runs in reverse of the logical order. In all cases, bracket the exposures, on the basis that film is cheap enough to buy in quantity to record these rare events. An action replay has never been known.

References

Allen C and Allen D, *Eclipse*, Allen and Unwin (1987)

Espenak F, *50-year Canon of Eclipses*, 1986–2035, NASA (1987)

Littman M and Willcox K, *Totality*, University of Hawaii Press (1994)

Meeus J, *Elements of Solar Eclipses*, Willman-Bell (1989)

Zirker J B, *Total Eclipses of the Sun*, Princeton University Press (1995)

Chapter 4

The Moon

Jeremy Cook

Why should one observe our Moon? What is it about that silvery globe that could induce anyone to spend hours on a dark night studying its surface? After all, in the last thirty years, spacecraft have been round the Moon, landed on it, photographed it – and men have walked on it and brought back many kilograms of its surface rocks to Earth for study. What is so special about it?

Despite those visits by spacecraft, there is still a large amount that is not known about the Moon, and in fact it is only very recently (1994) that the whole of the lunar surface, including the North and South polar regions, was comprehensively mapped. The visible hemisphere of the Moon, always turned towards the Earth, is shown in Figure 4.1.

Instruments

The choice of a telescope site for observing the Moon is not any different from that for any other astronomical object. A dark site is useful, but in this case it is not essential, since the reflectivity of the lunar surface and its nearness to the observer means that it can frequently be observed, even when light pollution would otherwise have brought an evening's observation to a halt.

It is up to the individual – and the individual's pocket – whether to use an observatory or a garden shed, or to observe at an open site with a portable instrument. There is an advantage to observing in some form of

Figure 4.1 The visible hemisphere of the Moon. South is at the top.

9 December 1992
2021 UT

shelter on a cold or breezy night, but much useful work can be done on an open site with either a telescope fitted to a fixed mounting, or one that is portable.

What form of telescope is the best for lunar observation? Refractors and the various forms of reflecting telescopes all have their adherents. Under good seeing conditions, the sharpness of lunar surface detail as seen through a refractor of 10-cm diameter OG or larger takes some beating. In light-gathering power, the nearest equivalent to the 10-cm refractor is the 15-cm diameter Newtonian reflector. Although, in general, reflecting telescopes cost less than the equivalent refractor, the obstruction to the centre of the optical path caused by the secondary mirror results in a small, but significant, loss of detail. However, in terms of cost, the reflecting telescope is much less expensive than its equivalent size refractor, and 30-cm and 45-cm Newtonian telescopes are not uncommon.

Whichever type of telescope one uses, it must be fitted to a stable mounting, preferably of the equatorial type. However, much useful work has been, and still is, carried out using either a straightforward altazimuth mount, or on what is known as a Dobsonian mounting. It is a fact that the larger the diameter of the object glass in a refractor, or the primary mirror in a reflector, the better the theoretical resolution of the image. That this is not necessarily true in practice depends on the column of air, moisture and dust that lies between the telescope and the outer layers of the atmosphere. This column of air is more or less unstable, and parts of it tend to move around causing distortion in the observed image. Under poor observing conditions, a 15-cm telescope can frequently out-perform a 30-cm telescope, and the 15-cm Newtonian reflecting telescope remains one of the most suitable telescopes for the beginner.

The type of eyepiece used also depends on the observer and his or her pocket. Purchase the highest quality eyepiece that you can afford; the Orthoscopic is a good, all-round performer. Focusing mounts usually take eyepieces which are fitted into a nominal 1.25-in (31-mm) diameter tube, with focal lengths ranging from 4-mm to 35-mm or more. Some mounts take larger diameter eyepieces, some smaller. The maximum usable magnification (or power) for lunar observation depends on the telescope and on the seeing conditions. Typical magnifications used are around 80× to 120× (magnification = focal length of mirror system or object glass, divided by focal length of eyepiece). Sometimes higher magnification may be used (up to about 450×) when seeing conditions are Antoniadi II or better.

The Observation

It is useful to have a map of the lunar surface available for reference, and versions of these are available in greater or lesser detail. Books which contain photographs of the lunar surface are also of great assistance in identifying features under differing illuminations which might otherwise be thought to have had temporary physical changes – known as Transient Lunar Phenomena (TLP). Nevertheless, any suspicious change in a feature which is thought to have occurred during an observation should be recorded in detail. These apparent changes, which mostly can be accounted for as purely visual effects, include temporary coloration in small areas (not to be confused with spurious colour due to refraction effects in the Earth's atmosphere), localised loss of detail, flashes and glows.

Observing aids include filters, and two of the most frequently used are the red Wratten No. 29 and the blue-green Wratten No. 44A. They may be used to highlight suspect colour changes in TLP work, while neutral density filters ranging from 0.1 to 4.0 in the form of an extinction photometer are used to measure relative brightness of features for similar work.

At the start of each observing session, it is useful to

have available a record book in which to make notes of the equipment and magnifications used, and sketches with notes of any features which are particularly under observation, and any peculiarities. A separate sketch pad can be used, making notes as you go along of the brightness or darkness of the various parts of a feature in a scale of 0 to 10 (0 darkest, 10 brightest). Draw at a generous scale – say 10 km to 1 cm (approximately 16 miles to 1 inch) so as not to be too cramped. Use a grade B pencil, and if the feature being drawn is close to the terminator, aim to complete your initial drawing within 15 minutes or so, as the shadows will alter perceptibly if more time is taken. As the illumination from the Moon is so bright, do not worry about destroying your dark adaptation by using a torch while you are drawing.

Having made the initial drawing at the telescope, with added notes on the relative brightness and darkness of its features, the drawing can be brought indoors and used as a basis for a final drawing in pencil or ink. How to achieve subtle light and dark tones on the final drawing is up to the observer. Several methods are currently being used, with pencil shading or the 'dotting' method being preferred.

Of the data that should be recorded along with each drawing, three items are very important: date; time of start and finish of drawing; and seeing conditions. Date and time should present little difficulty with the latter being expressed in Universal Time (UT), but note that Universal Time is one hour less than British Summer Time when that is in operation. If observations are carried out near to midnight, the date and time should be expressed using the double date, e.g. 1994 March 4/5 23:04 – 23:17UT. Seeing conditions should be recorded on the widely-used Antoniadi scale, which ranges from I corresponding to 'perfect seeing without a quiver', through II, III and IV, to V, which is 'very bad seeing, not allowing even an adequate sketch to be made'. The transparency should also be noted as a measure of the clarity of the atmosphere, and can either be recorded as the faintest star visible near the celestial pole, or more conveniently as 'good', 'moderate' or 'poor'.

Although the Moon is in a captured rotation about the Earth and keeps approximately the same hemisphere turned towards it, there remain two small motions which give rise to librations in longitude and latitude. These are caused respectively by the slight ellipticity of the Moon's orbit, and the inclination of the Moon's axis to its orbital plane. The result is that we can see about 59% of the lunar surface as the Moon wobbles or librates about ±6.5° North–South and ±7.5° East–West. Figure 4.2 plots the course of these librations over a period of six years.

The librations in these two directions are not synchronised, and so different features on the lunar limb are successively presented for observation. Due to the

Figure 4.2 Libration – variation in the mean centre of the Moon's disc relative to the centre of the visible surface, plotted here over a period of six years.

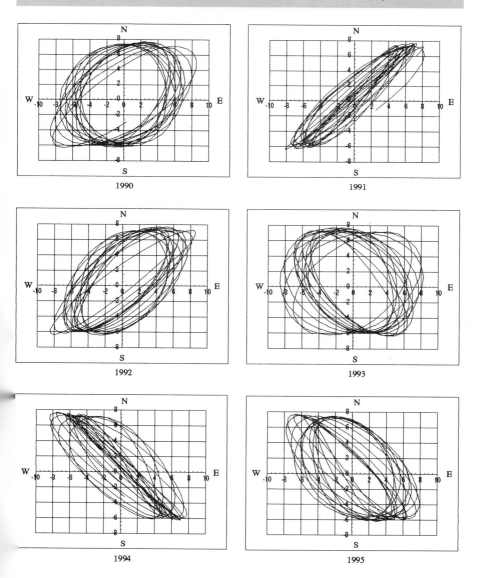

Figure 4.3
Detail visible on the extremes of the Moon's limb. The Mare Crisium is well shown.

4 April 1977

fact that the Moon is relatively close to the Earth (384,000 km) the amount that can be observed of the libration regions varies according to the position of the observer on the surface of the Earth. Figure 4.3 shows detail that can be observed on the extremes of the Moon's limb. Computer programs exist for the amateur to calculate the libration direction and amount for a given date, and for much other data relating to the Moon.

For over 200 years, the classical way of expressing longitude on the Moon had been with Mare Crisium in the West and Grimaldi in the East, which corresponded to directions in the Earth's sky. However, in anticipation of manned landings on the Moon and the requirement that sunrise on the Moon should be with the Sun in the

Table 4.1. Selenographic colongitude

Moon's phase	Longitude of morning terminator (selenographic colongitude)	Longitude of sub-solar point	Longitude of evening terminator
New Moon	90°E (270°W)	180°W	90°W
First quarter	0°	90°W (270°E)	180°W
Full Moon	90°W	0°	90°E (270°W)
Last quarter	180°W	90°W	0°

familiar East, and sunset with the Sun in the West, the directions were changed in 1961 by the International Astronomical Union (IAU). After the change, when considering selenographic locations or directions on the Moon, Mare Crisium is referred to as being in the East and Grimaldi in the West. Longitudes are sometimes quoted as running from 0° to 360° in a westerly direction, or more often as running from 0° to 180°W/180°E back to 0° again.

Some scientists, mostly those from the geophysics community, still refer to longitude as being positive in an anticlockwise direction if one were to look down upon the Moon's North pole. However, when referring to drawings or photographs, the IAU system should be used, with 'E' or 'W' added to longitudes in case of doubt.

An additional piece of information that should be added to lunar drawings is the Sun's selenographic colongitude, i.e. the longitude (IAU) of the morning terminator. The position on the lunar surface where the Sun is directly overhead is called the sub-solar point. The morning terminator is 90° to the West of this, while the evening terminator is 90° to its East, as shown in Table 4.1.

Details of the colongitude on specific dates can usually be obtained from tables in the yearly almanacs issued by major observatories and national astronomical societies. These are accurate enough for most purposes.

Some of the most spectacular views of the Moon's surface are obtained near the terminator, when the low Sun angle causes shadows to exaggerate the depth of shallow features (see Figure 4.4) and to produce elongated shadows of mountain peaks. A typical example of the latter is the needle-like shadow caused by a peak on the Eastern rim of Plato, which stretches across the crater just after local sunrise and which retreats across

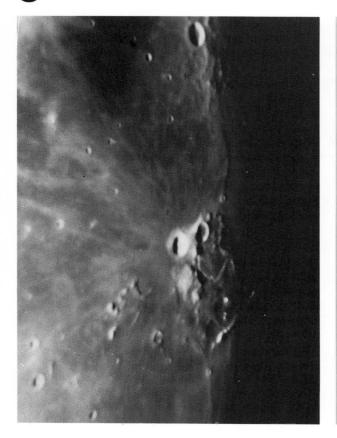

Figure 4.4
Shadows thrown by the Sun cause features near the terminator on the Moon's surface to be dramatically exaggerated. The two craters near the centre are Aristarchus and Herodotus.

27 March 1991 2010 UT; 8-inch Maksutov, ¹/₂-inch eyepiece, f/80

the crater floor rapidly at first and then ever more slowly as local noon approaches. Although lunar craters are in general circular, their appearance to the observer depends very much on their position on the Earthward part of the spherical Moon. Those craters that are within 20° or so of the centre of the visible disc do appear essentially circular, but those further away appear more and more elliptical the nearer they approach the limb. Any crater away from the central zone will alter its appearance to a greater or lesser extent depending on its position, the amount and direction of the libration, and the lighting conditions. This is why the appearance of a feature is not exactly repeated every lunar month but every 18 years 10.25 days (the so-called Saros period), when the Sun, Moon and Earth are very close to being in the same relative positions. However, once or twice a year conditions are repeated sufficiently closely for many purposes except the most critical observations.

The tilt of the Earth's axis not only causes the changing seasons but also means that the plane of the ecliptic is at its highest in winter after dark and therefore the Moon's elevation is then at its most favourable for observation. In the summer months some useful observing can be achieved, but this is a good time to ensure that equipment is put into proper order ready for the observing season to come.

The Earth rotates on its axis relative to the Sun in 24 hours, but as the Earth itself is in orbit around the Sun, it appears to rotate in approximately 23 hours and 56 minutes relative to the much more distant background of stars. Relative to an observer on the surface of the Earth, the background of stars therefore appears to rotate in an East–West direction in the same 23 hours and 56 minutes. The Moon also appears to the observer to move in the same direction, but since it is rotating around the Earth in the same direction, but slightly faster than the Earth spins on its axis, it seems to travel Eastwards through the star background at the rate of about one Moon's breadth every hour. This is the reason why telescopes need to be driven faster than sidereal rate to track the Moon. Of course the movement of the Moon is somewhat more complex than this and, in practice, telescope drives have to be adjusted in both declination and right ascension during long observing periods.

Occultations

One form of lunar observing is that of occultation timing, determining the moment when an unilluminated part of the Moon places itself between a star and the observer – known as an immersion. As we have seen, the Moon moves in an Easterly direction in the sky against the background of stars, and will inevitably pass in front of a particular star at some time, provided that star is within a band of about ±20° of the ecliptic. The apparent motion of the Moon is quite complex so that stars up to about 29° from the ecliptic may occasionally be occulted. Figure 4.5 shows the near-occultation of the star Aldebaran.

The value of these occultation timings is that they assist in the refinement of the positions of the occulted stars relative to the Moon. Of course, now that the spacecraft *Hipparcos* has completed its mapping of stellar positions, the process can be reversed and the position of the Moon can be further refined.

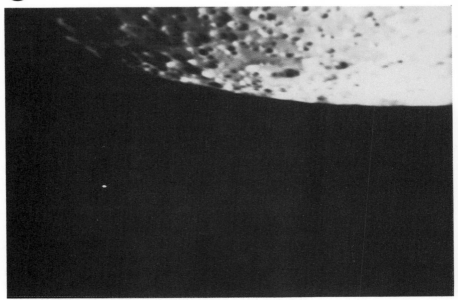

Graze occultations occur when the star being occulted just slides along the limb of the Moon. It may be occulted momentarily perhaps several times as it is hidden behind high ground. By the timing of individuals in a team of observers set up in a line across the track of such an event, the profile of the lunar limb at that position can be determined so that any corrections can be applied to the series of Watts lunar profile charts.

Another form of occultation is the eclipse of the Sun by the Moon – the solar eclipse; and that of the Sun by the Earth – the lunar eclipse. Of these, the most spectacular (but the shortest in duration) is the solar eclipse, the width of the shadow on the surface of the Earth being used to calculate the angular diameter of the Sun. The lunar eclipse lasts longer and may be seen over a much larger area, and the timing of the umbra as it reaches specified features on the lunar surface has been used to revise our knowledge of the shape of the Moon.

Figure 4.5 The near-occultation of the star Aldebaran.

12 February 1981
2316 UT

Recording the Observation

The best method for an observer to learn about the topography of the lunar surface is to concentrate on particular features, drawing them under a number of different angles of illumination. By collecting together drawings of the same feature in an observing book, the apparent changes in their shapes with angle of illumination will be shown. Also, with some practice, the observer will start to see more and more detail on the lunar surface as the eye and brain learn together. There is no need to worry if you do not initially have great confidence in your drawing ability. This quite definitely improves with practice, and anyway you are carrying out this exercise for your own benefit.

Other means of recording observations include photography, video recording, and CCD imaging (CCD: charge-coupled device – an array of electronic sensors which respond to changing levels of incident light). The simplest method of using a camera to take pictures of the Moon through a telescope is to focus the telescope by eye, focus the camera at infinity, hold the camera lens to the eyepiece and press the shutter. This may result in a suitable picture, depending on how well the telescope and camera were focused, the image magnification, the film speed, and whether the telescope and camera were inadvertently shaken at the moment of operating the shutter. Expect about one in twenty of these snapshot exposures to be satisfactory.

As a better alternative, a single-lens reflex (SLR) camera without its lens may be used. The camera's body is attached to the eyepiece mounting so that the image of the Moon is focused directly on to the film plane. The image can be observed using the reflex viewfinder in the camera, and correct focus obtained. This is a much improved method, since the image can be seen through the camera's viewfinder up to the moment of exposure, and the camera's exposure meter can be used as a guide to the shutter speed to be used.

Now that video cameras are becoming considerably smaller and lighter, it is possible to use them in a manner similar to that described for still cameras, although if mounting one directly on the telescope, account must be taken of its larger mass on the balance of the telescope. By replaying a satisfactory recording and using the freeze-frame facility, advantage can be taken of displaying moments of good seeing, although the images may tend to be rather more grainy than one is used to.

CCDs can also be used to obtain images of the lunar surface and they are usually mounted in a manner similar to the SLR camera, with the image being observed for focusing on the associated computer monitor. Two disadvantages of using CCDs compared with photographic film are that at present they do not give as good definition and, as the arrays themselves are small in size (about 4 mm × 6 mm compared with 24 mm × 36 mm for photographic film), the area of the lunar surface that can be covered at a given magnification is reduced to about 1/36th. However, among the advantages of CCDs are that they are very sensitive and therefore allow short exposures, they have a linear response and can be used photometrically after calibration, and since they interface directly with a computer, the resulting images can easily be computer processed and enhanced without waiting for chemical processing. Larger CCD sensors are being developed with better resolution, but it is likely to be some time before they become readily available to the amateur together with the improved computing power that will also be required.

References

Hatfield H, *The Amateur Astronomer's Photographic Lunar Atlas*, Lutterworth Press (1968). (This invaluable Atlas is unfortunately out of print, but a reduced version is given in the BAA booklet *Guide to Observers of the Moon*, edited by Patrick Moore.)

Brice F W, *The Moon Observer's Handbook*, Cambridge University Press (1988)

Rükl A, *Atlas of the Moon*, Hamlyn (1990)

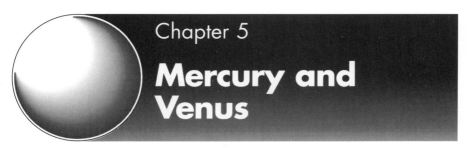

Chapter 5

Mercury and Venus

Richard M. Baum

'Both Mercury and Venus are troublesome objects for observation,' cautioned Patrick Moore in 1963. This must ever be so where the Earth-based observer is concerned, since the difficulties created by phase are compounded by those peculiar to each object. Nevertheless, the situation is not quite what it used to be. At least we can now relate visual experience to the truth of these words, and that in itself makes the two planets more attractive as objects of telescopic scrutiny.

Mercury

Close proximity to the Sun renders Mercury difficult to observe. At maximum elongation it is never more than 28° from the Sun, and is only observable in daylight and bright twilight. But in daylight the glare from the sky reduces contrast on the tiny disc, while at twilight the planet is low in the sky, and is seen through the murky, disturbed air near the Earth's horizon. The best time to observe, then, has to be determined according to local conditions.

Telescopically, this small, moon-like world demands apertures in excess of 15 cm, although smaller instruments have yielded interesting results. That, however, is more the exception. What can the observer expect to see? If conditions are favourable the disc will resemble a low-resolution view of the Moon, with a mottling of dark and light spots. The latter may correspond to the bright ray systems that pepper the surface of the planet.

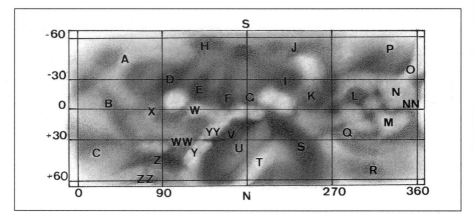

The existing albedo cartography of Mercury is excellent, but its reliability is variable, especially around longitudes 0°–100° and 190°–200°, the latter zone being in the area not covered by *Mariner 10*. Moreover, it is chiefly based on observations made before the revision of the rotation period. There is thus some justification for producing a new albedo chart of Mercury, such as the one shown in Figure 5.1, based on new visual and photographic data.

Observers familiar with Mercury know that its elongations from the Sun (of which there are about six a year – three in the morning sky and three in the evening) are brief and differ considerably in favourability. Those considered favourable occur when Mercury is high above the horizon during twilight (morning apparitions in the Autumn, evening ones in the Spring). But the planet is often further from the Sun (though low on the horizon at twilight) during so-called unfavourable elongations, as seen by northern observers, and fails to attract interest. To get good coverage sufficient to construct a complete map one must attempt to observe during all apparitions, not just the most favourable. In addition, best coverage will only be obtained if the planet is observed throughout each elongation, not just the few days around the greatest elongation, when Mercury is half full.

Detection of surface markings on Mercury depends critically on many factors; these include the aperture of the telescope, magnification, absorption in the telescope, seeing conditions, transparency and brightness of the sky, Mercury's phase and orbital distance from the Sun, and the inherent contrast and size of surface markings.

Figure 5.1 An albedo chart of Mercury, based on new visual and photographic data.

Venus

Observation of Venus depends most critically on telescopic and seeing conditions that produce optimum contrast, rather than telescopic resolution. True understanding of this very important, though often overlooked, fact is crucial to effective observation of the planet, the third brightest object in the sky and, after the Moon, our nearest neighbour in space.

Lomonosov (1761) and Rittenhouse (1769) proved that Venus has an extensive atmosphere. H. N. Russell (1899) derived some quantitative data about its optical properties from visual observations of his own, and from earlier work by Schröter, Herschel, Mädler, Lyman and others. But at that point visual achievement stopped, and Camille Flammarion observed: 'The most careful discussion of all the observations leads us to think that the grey spots now and again seen upon Venus are the effects of contrast due to solar illumination, and that the less definite shadings are of an atmospheric nature, incapable of furnishing us with any serious data as to the rotation of the planet. We may state here more than ever that every man sees after his own fashion and draws after his own fashion. We are sure of nothing.' (Flammarion C, *Knowledge* 20 (1897).)

In 1927, F. E. Ross photographed the planet in the ultra-violet. His findings unexpectedly vindicated earlier reports of brighter polar spots and dusky banding, but were not followed up until 1957 when C. Boyer repeated the experiment, and discovered that the UV markings reappeared every four days. Not surprisingly, they have a windswept look. They are located at an altitude of about 65 km above the surface, and are composed of sulphur crystals and sulphuric acid droplets.

Today, thanks to a flotilla of space probes, radar, and sophisticated ground-based techniques, Venus is no longer a twilight mystery. But the visual observer is still faced by the old problem, the cloud blanket. So has nothing changed? Of course it has, though many fail to grasp the fact. For now we can observe with an informed eye. We can view old problems in a different light, and perceive new opportunities. Take the case of the polar brightenings. Until Ross proved otherwise in 1927, they were thought to be illusory. Yet all the time we were observing the cloud swirls of the great polar vortices, and the associated ring of cold polar air, first described by Percival Lowell in the late 1890s as a dark collar to the bright hoods. There are other correlations, too, enough to satisfy us that visual observation can yield meaningful results. We may quote the example of H. Schwabe. What began as a search for intra-Mercurial planets caused him to more closely examine the Sun, and to make daily counts of its spots. To some the tabulations were amusing asides, the jottings of a backyard stargazer. But derision evaporated when a very important truth was disclosed, the 11-year solar cycle.

This is not to say that visual observation will lead to results of similar magnitude; rather, to remind us that Venus, like Mars and Jupiter, is a dynamic planet, and must be observed as consistently as possible over as much of its orbit as circumstances allow, not just at the most favourable intervals when it is easily accessible. This creates problems, not impossibilities. Always remember, though, that the markings are exceedingly faint and elusive – so much so that their existence is ever in doubt. At such times the eye is apt to play tricks, and the observer must maintain a constant guard against optical illusions.

The observer should attempt to use the same telescope, at the same aperture and the same magnification, from the same location at about the same time each day, conditions permitting. The quality of seeing, transparency and sky brightness should be logged.

Instruments, Strategy and Techniques

Instruments

I do not propose to revive time-worn arguments about what aperture to use, or indeed what magnification to employ. All depends on who stands at the ocular; Gruithuisen discovered the polar hoods with a tiny refractor, barely two inches in aperture. What does count is experience, a critical eye, sober judgement, local conditions and clean optics. The most important area of research open to the amateur is UV photography. UV filters such as Schott UV5 and UG2 have maximum transmission between 320 and 370 nm, because ozone in the Earth's atmosphere absorbs incoming light at wavelengths shorter than 320 nm, and the Venus UV markings tend to fade at wavelengths longer than 370 nm. There is also scope for near infra-red work and CCDs.

Methods

Here I shall confine comment to direct visual study, and
then only in the broadest manner. A quick glimpse by
averted vision is the best way of picking up faint markings. The longer one looks the more elusive they
become. There is evidence to suggest that some
observers may be unusually sensitive to light of short
wavelength, implying that others may be similarly
endowed in the opposite direction.

Most observations are made through disturbed air
from stations which are situated near or within the
Earth's terminator. At a western elongation, image
degradation caused by agitation occurs as the rotation
of the Earth plunges the still, cold night air into the heat
of the day. After the Sun has risen, the warmed air
becomes more steady, and the image perfects just as the
observer is obliged to pack up and leave for work! At an
eastern elongation, images are relatively still during the
afternoon. I find the best moments for observation
occur about two to three hours before sunset, extending
a short time into twilight. In 1991, excellent seeing was
experienced about 30 minutes before sunset. The Revd
T. E. R. Phillips found conditions better in the morning
and evening than at midday. The problem is geographical. What is applicable to one observer is not necessarily so to another.

Intensity Estimates

Techniques of visual photometry involve estimating relative intensities of markings seen on Venus. These are
initially made in white light. Instead of verbal descrip-
tions, most observers use a numerical scale to describe
what they see on the disc of the planet. The following is
the scale devised by Patrick Moore for use by the
Mercury and Venus Section of the British Astronomical
Association (BAA).

0   Extremely bright spots, polar hoods, limb
    brightening.
1   Bright spots.
2   General hue of the disc.
3   Shadings dark enough to be seen with certainty.
4   More definite shadings, sometimes seen near the
    terminator, and adjacent to the polar hoods.
5   Unusually dark shadings, exceptional.

Phase Estimates

The most accurate results are derived from micrometric measurement, or measures of photographic negatives. Some observers estimate by comparison of the telescopic image with a set of pre-dawn phase blanks. These are held at arm's length. One then checks the telescopic image of Venus against the blank that most nearly corresponds to what is visible, and draws accordingly. Wratten filter No. 15 (yellow) is recommended to reduce glare when making visual estimates.

Filters

These are used to establish how the planet looks in light of different wavelengths. Wratten No. 15 (yellow), as mentioned earlier, has been found useful in reducing glare. This is standard and can be used with all apertures. However, filters vary in density and some are unsuitable for small apertures. For instance Wratten No. 29 (red) and Nos 47 or 49 (blue) are too dense, and Nos 25 (red) and 44A (blue) are recommended for small instruments. The planet is better defined in a blue sky if viewed in red or yellow light.

Drawings

Like photographs, visual renditions are meant to be data resources. We know what Venus looks like, that its markings are faint and fugitive, and that sometimes nothing at all can be seen. What is required is information and this can be conveyed in annotated diagrammatic form. Finished drawings are matters of personal preference. Lack of artistic ability need not deter the keen observer if serious study is the intent.

Subjects for Scrutiny

Bright and Dusky Markings

Venus is permanently covered by clouds and is characteristically featureless in the visible wavelengths of light, but exhibits variable albedo markings in the near ultraviolet. Of these, the most conspicuous is a planetary-

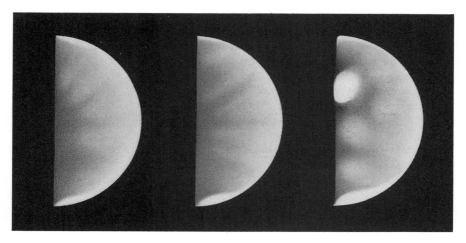

Figure 5.2
Telescopic drawings of Venus showing polar hoods, dusky markings, and a localised bright region near the terminator.

Left: 3 November 1991 0800–0850 UT; 115-mm OG ×186
Middle: 4 November 1991 0800–0850 UT; 115-mm OG ×186
Right: 5 November 1991 0828–0920 UT; 115-mm OG ×186

scale, dark Y-shaped feature. This was defined in Earth-derived and *Mariner 10* images. It rotates about the planet with a period of 4–5 days, but changes dramatically from one appearance to the next, and occasionally disappears for several cycles. A dark, reversed C-shaped or bow-like marking shows a similar pattern of activity. Both lie on or near the equator, and their trailing edges follow in the rotational direction. These global and small-scale markings supply important information on horizontal and vertical cloud structure, atmospheric waves and wind speeds at the cloud tops. Long-term evolution of cloud properties and atmospheric dynamics, as well as small-scale changes in cloud morphology, are possible areas of study for the well equipped observer who contemplates UV photography of the planet.

At rare intervals, localised bright areas appear (see Figure 5.2). These are independent of the bright polar hoods described below. Very rarely the disc seems to be mottled by an intricate network of spots and streaks. More exotic are the needle-like specks of light in and around the polar hoods which were extensively reported by Barnard, McEwen, Molesworth, Trouvelot, and other classic observers.

The Polar Hoods

The geometry of an isotropically scattering sphere will cause the poles to appear darker than the equatorial region. Hence the bright spots or hoods at the poles of Venus are anomalous. They were discovered by the German astronomer Franz von Paula Gruithuisen in

December 1813 as he examined the planet with a small spyglass. But until photographed in the ultra-violet by F. E. Ross in 1927, they were thought to be illusory. Now they are known to correspond to the bright polar cloud vortices imaged by the spacecraft *Pioneer Venus*.

In general, the polar hoods exhibit a two-part structure, consisting of a bright polar band, situated between 45° and 65° latitude, and a slightly darker cap at higher latitudes. For 50% of the time, one or more narrow dark polar bands separate the bright polar band from the cap. The absence of small-scale features invests the polar regions with a typically smooth appearance.

Sometimes one cap is visible, the other not. At other times both hoods are absent. Indeed the configuration in which the two polar regions appear bright is neither typical nor permanent. There is strong evidence to suggest the hoods and their associated dusky collars fluctuate in visibility and size, and brighten on time-scales of between a month and a year, relative to low and mid-latitudes, but do not show any correlations between their brightness variations.

The hoods are not always placed exactly at the cusps. Occasionally they appear as narrow, bright bands along the limb, distinguishable only from the limb brightness by their snow-like whiteness. At other times they change into elliptical bright patches or ill-defined white spots which merge into the general brightness of the disc.

The dark collars adjacent to the polar hoods may be the visible signs of the polar ring. This is a current of cold air, almost 1000 km in diameter but only about 10 km thick, encircling the pole at a radial distance of approximately 2500 km.

Long-term visual and photographic monitoring of the hoods and their dark collars is available to the dedicated visual observer.

Contour Anomaly

Ideally the boundary between the sunlit and dark hemispheres of Venus should appear as a perfectly smooth half ellipse, symmetrical with the apparent equator. But often its length is distorted by large-scale deformations. These may be long, low swellings, interspersed with shallow depressions, or flat segments. Again, one half of the curvature may appear to be concave, the other convex. Small-scale deformities are often reported.

While some of these appearances may be due to variations in the height of the cloud deck, most seem to be contrast effects. A dusky marking near the terminator can give the impression of an indentation. Conversely, localised brightness in the same region will give rise to a small 'hump' (see Figure 5.2). Similarly, irradiation will cause a bright polar hood, or localised brightness near the limb, to project above the apparent periphery of the telescopic image. A small, bright limb projection was seen and photographed by J. D. Greenwood in March 1985.

Asymmetry at the cusps in which one cusp regularly appears sharp and tapered, the other blunt, is well attested but difficult to explain. It may represent very large relief in the cloud cover. Systematic observation may enable us to decide if it is real or not.

Phase Anomaly

This is better known as the Schröter Effect, a term introduced by Patrick Moore in 1955 after the founder of modern comparative lunar and planetary study, J. H. Schröter, who was the first astronomer to call attention to the difference in date between the observed and theoretical dichotomy (half phase).

Predicted phase is the ratio of the area of the illuminated part of the apparent disc to the entire disc regarded as circular. It is calculated from the orbital geometry of Venus and the Earth, but does not always correspond with the observed phase. The difference is especially noticeable at dichotomy, when the observed event can be up to nine days early at an evening elongation, and almost seven days late at a morning elongation.

It has also been established, both visually and photographically, that the phase in red light is greater than in blue. Scattering of blue light at the terminator may account for this. Also the difference in phase (red to blue) may vary. For this work, Wratten filters No. 25 (red) and No. 44A (blue) should be used with small apertures, and the more dense No. 29 (red) and No. 49 (blue–violet) with large telescopes.

From experiments with 'model planets' in 1971 the Russian astronomer V. A. Bronshten concluded that visual estimates of phase could distort the picture. He found there is a tendency for observers to underestimate at the gibbous phase, and to exaggerate when Venus is a thin crescent. There is no definitive explanation of the anomaly. It may be part physiological, part psychological; but that is not the whole story. Variations in cloud height due to climatic and other causes could materially affect the shadow profile at the terminator, but – as with the Ashen Light – the accumulation of observations has done nothing to elucidate the mystery.

Visibility of the Dark Side

The Ashen Light

A glimmer of greyish light on the dark side of the Venus was initially sighted in 1643, but first accurately described in 1714 by the English prelate William Derham who compared it to Earthshine. The light, which is very elusive, is conspicuous during the planet's passage through inferior conjunction. It is usually depicted as a uniform glow, though patchiness, mottling, dark limb brightness and other appearances are not unknown.

This Secondary, or Ashen, Light as it is called is one of the oldest unsolved enigmas in the history of planetary exploration, and probably the most controversial. Of course, if occurrence was established by weight of observation, there would be no problem. But circumstances dictate otherwise, and cause many to question its physical reality.

As we have seen, Derham likened the Ashen Light to Earthshine. But Venus has no moon. Nor can Earthlight be invoked as a cause. Nevertheless the analogy is apt, simply because the visibility of both Earthshine and the Ashen Light is phase dependent. But there is a significant difference; we know Earthshine is brighter than the sky. Even so, it is not normally visible in daylight, or easy to recognise in very bright twilight. Further, it is often impossible to decide if the unlit parts of the Moon are brighter or darker than the sky. Also, Earthshine is less easy to detect when more than 40% of the Moon's visible surface is sunlit.

This suggests that nocturnal brightening on Venus, if it exists, will be seen only at large phase angles and when the planet is viewed in a dark sky. The most favourable time to look is around the start of astronomical twilight at an evening elongation, and before its end in the morning sky. But Venus is then at a low altitude and image quality is impaired by bad seeing, poor transparency, and atmospheric dispersion. The observer also has to contend with the effects of secondary spectrum, contrast-induced illusions, and glare from the sunlit part of the planet. Of course the latter can be overcome by the use of an occulting bar kinked into the shape of the crescent and placed in the ocular. A search for the Ashen Light may thus be conducted while the sunlit side of the planet is hidden by the bar. Even so, a positive identification is not guaranteed. Some authorities recommend Wratten No. 35 (purple) and No. 58 (green) filters but, again, results are unreliable.

With no general agreement as to whether the phenomenon is real or false, and few trustworthy observations on record, the intending observer may decide the subject is not worth pursuing. This would be a mistake. The difficulties involved should sharpen the mind and challenge the ingenuity. The fact that there is a small core of confirmed reports is sufficient reason to undertake an investigation. But considerable care will need to be exercised.

Of the hypotheses advanced in explanation of the phenomenon, only one shows any promise. This is based on a series of infra-red photographs of the dark side taken by D. A. Allen and J. Crawford in 1983 which revealed what are thought to be high concentrations of sulphuric acid in the lower clouds at an altitude of about 50 km, just below the UV features. These cloud patterns rotate in 5.4 days, and may be illuminated by atmospheric scattering of solar radiation from the sunlit side of the planet. Could these dimly lit clouds occasionally become visible in white light, thereby giving rise to the ghostly gleam of the Ashen Light?

No doubt exists that the Ashen Light, as we know it, is an *ignis fatuus*, a mystery of mysteries that will continue to excite speculation until its truth is unveiled. Whether this will be achieved by techniques other than those traditionally deployed remains to be determined.

The Dark Phase

D. F. J. Arago referred to this as 'negative visibility'. As the phrase implies, it relates to all those cases in which the unilluminated side is seen as darker than the surrounding sky. However, there is no physical process that seems capable of explaining such a phenomenon and it may be regarded as an effect produced in the eye by the sharp contrast in brightness between Venus and the sky. It is more noticeable if observed with small apertures at low magnification, and may also be seen during the gibbous phase. The effect can be simulated if a crescent cut from translucent paper is placed in front of a plate of ground glass and then strongly illuminated from behind. Viewed at a distance with the naked eye an illusory dark area adjacent to the terminator will be seen. If the crescent is very thin, the eye will tend to complete the illusion and give the impression of the unlit part appearing darker than the surrounding background.

Significantly, daylight visibility of the dark side necessarily implies high intrinsic brightness. But this is not the case. There are no reports of the Ashen Light being this conspicuous, only accounts of something very elusive at the threshold of vision.

An interesting possibility worth following up was put forward during the last century by W. T. Lynn and W. Noble. They proposed that the dark phase might be explained by projection of the planet on the solar corona. The optimum time to search occurs a week or so either side of inferior conjunction but, as already mentioned, it may prove difficult to determine if the dark side is brighter or darker than a daylit or bright twilit sky, or even if it is projected on the solar corona. Careful studies of Earthshine on the Moon, as seen by daylight and in bright twilight, might shed some light on this vexed problem.

Observation of the Dark Side

It is important to establish the existence and nature of the dark side phenomenon. An occulting bar fitted in the ocular is essential, and experimentation with filters is desirable; Wratten No. 58 (green) and No. 35 (purple) are recommended. Photography should also be attempted. Try different magnifications and telescopes to test for illusion. Note precisely the condition and brightness of the surrounding sky. Always keep an accurate record, omit nothing, and remember that negative observations are statistically important.

The Twilight Arc and Extended Cusps

As Venus approaches to within 15° of the Sun, tenuous threads of blue-white light edge out from the needle-like cusps to extend themselves around the dark limb (see

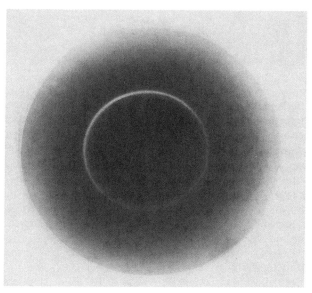

Figure 5.3 Cusps of light around the dark limb of Venus.

2 April 1977 1840 UT; 115-mm OG ×186; excellent seeing and transparency conditions

Figure 5.3), until finally, several days before inferior conjunction, they meet to encircle the planet with a fragile ring of silvery light. This is the twilight arc, and is due to sunlight reflected in the upper atmosphere of the planet. It is a spectacle of rare and delicate beauty and is easily visible through a small telescope. However, this is an observation *which must be carried out with the utmost caution, because of the close proximity of the Sun,* and should not be attempted by the novice. Accurate alignment of the equatorial is essential and correct use of setting circles is mandatory; also, ensure that the finder telescope is capped, and fit a long dew cap or sunshade.

The complete Twilight Ring was first seen by Guthrie in 1842. But J. H. Schröter in 1790 had already noticed that the thin crescent of Venus near the Sun could be seen extended beyond a semicircle. H. N. Russell observed the complete ring in 1899. He also correctly interpreted the appearance, and developed the geometric theory of cusp extension and computed the maximum height of the scattering agency above the opaque cloud stratum. He calculated the angular extent of the bright cusp extensions, i.e. those commonly observed through the haze which is prevalent at most stations, to be about 68 seconds of arc, and that of the longer, more tenuous arcs, visible in excellent seeing, 3° 21′. The first photographs of the phenomenon were taken at the Lowell Observatory in November 1938.

Transits of Mercury and Venus

Since their orbits are slightly inclined to the ecliptic, both planets usually pass north or south of the Sun at inferior conjunction; but if conjunction occurs when either planet is near its node, then from the Earth it is seen as a small, black spot moving East to West across the Sun's disc along a path sensibly parallel to the ecliptic. The special interest of transits is their availability for the purpose of finding the parallax and distance of the Sun, as first pointed out by Halley in 1679. However, other methods of achieving this have since been found to be more trustworthy.

Mercury will transit on 1999 November 15, 2003 May 7 and 2006 November 8. Venus transits are very rare and occur in pairs: the last pair were in 1874 and 1882, the next pair are due 2004 June 8 and 2012 June 6. Residents of the British Isles will see the whole of the 2004 transit, but the transit due in 2012 will already be in progress when the sun rises.

Mercury and Venus pose unique problems for the Earthbound observer. With the former we have elusiveness; with the latter, intractability. Nevertheless, the frustration inherent in their observation ought not to deter the serious worker. Quite apart from the æsthetics, it is useful to revisit what has already been achieved, simply as an exercise in self-education. If this process is intelligently conducted then the unexpected may be sprung, and a new excitement generated. With Venus we confront the eternal problem of the clouds. But, as I said at the beginning of this chapter, the planet is now observed with an informed eye, and some of the uncertainty has been removed. The Ashen Light is still a mystery, and the suspected variations in the visibility of the polar hoods have yet to be resolved. Nor must we overlook the potential of the competent, well-equipped amateur in the field of UV photography.

Bibliography

Books

Cattermole P, *Venus*, London, 1994

Hunt G E and Moore P, *The Planet Venus*, London, 1982

Kuiper G P and Middlehurst B M *The Solar System. Volume III: Planets and Satellites*, Chicago, 1961 (especially Chapter 15, '*Pioneer Venus*', reprinted from the *Journal of Geophysical Research*, 85:A13 (December 30 1980))

Moore P, *New Guide to the Planets*, London, 1993

Papers

Bronshten V A, 'The Schröter effect in the USSR', *Journal BAA*, 81 (1971), 181–5

Curtis A C, ' "Ashen Light" observations of Venus, and a special occulting bar', *Journal BAA* 74 (1964), 229–34

Dollfus A, 'Étude visuelle et photographique de l'atmosphère de Vénus', *L'Astronomie 69 année* (November 1955), 413–25

Dollfus A, 'Observation visuelle et photographique des planètes Mercure et Vénus à l'Observatoire du Pic du Midi', *L'Astronomie 67 année* (February 1953), 61–75

Dollfus A, 'Venus: Evolution of the Upper Atmospheric Clouds', *Journal of Atmospheric Physics* 32:6 (June 1975), 1060–1070

Hiscott J, 'Ultraviolet Observations of Venus in 1969', *Journal BAA* 82 (1972), 198–9

Robinson J H, 'Report on the Observation of the Planet Venus 1956–1972', *Memoirs of the BAA* 41 (December 1974)

Ross F E, 'Photographs of Venus', *Astrophysical Journal* LXVIII (1928), 57–92

Whitaker E A, 'Visual observations of Venus in the UV', *Journal BAA* 99 (1989), 296–7

Chapter 6

Mars

Patrick Moore

It has been said that Mars is a very easy planet to see, but a very difficult planet to observe. There is a good deal of truth in this. Despite its small size, Mars at its best can outshine every other natural body in the sky apart from the Sun, the Moon and Venus; its magnitude can reach −2.8. On the other hand, when badly placed the magnitude can sink to 1.8, not much brighter than the Pole Star. The apparent diameter is also very variable; at maximum it is 25.7 seconds of arc, at minimum only 3.5 seconds of arc – about the same as that of distant Uranus.

This means that telescopes of even small or moderate size will show little on Mars except when the planet is reasonably close to opposition, and to complicate matters still further not all oppositions are equally favourable. The Martian orbit is much less circular than that of the Earth; the distance from the Sun ranges between 249,000,000 km at aphelion and 207,000,000 km at perihelion. The sidereal period is 687 days. Minimum distance from the Earth occurs at perihelic opposition, when the distance may be reduced to as little as 56,200,000 km, as in 1956 (even so, Mars can never come as close to us as Venus does). Figure 6.1 shows Mars in opposition in 1988, when the distance was 58,400,000 km. At aphelic oppositions the minimum distance never becomes less than 101,300,000 km, as in 1980. Opposition details for the ten years until 2005 are given in Table 6.1.

The mean synodic period is 780 days, so that Mars comes to opposition only in alternate years. Obviously, then, opportunities for really useful work are limited,

Figure 6.1
Mars in opposition, in 1988.

28 September 1988 2220 UT; $\omega = 195°.4$: $\delta = 23".6$: $\phi = -21°$, 115-mm OG ×186; seeing II (Antoniadi Scale)

and there are long spells when the would-be Martian observer must turn his attention elsewhere.

There is another point to be borne in mind. The axial inclination of Mars is 23° 59′, almost exactly the same as that of the Earth; southern summer occurs at perihelion. It follows that when Mars is best placed for observation from Earth, it is the southern hemisphere which is best displayed, and we never have as good a view of the northern part of Mars. In fact, there is a definite difference between the two hemispheres. The Mariner and Viking spacecraft have shown us that, contrary to our previous beliefs, Mars is mountainous and cratered; the southern part of the planet contains more craters, and is loftier, while the north is smoother and lower (though it is true that the two deepest basins, Hellas and Argyre,

Table 6.1. Oppositions of Mars, 1995–2005

Date	Closest approach to Earth	Minimum distance, km × 10⁶	Apparent diameter, arcmins	Mag.	Constellation
1995 Feb 12	1995 Feb 11	101	13.8	−1.0	Leo
1997 Mar 17	1997 Mar 20	99	14.2	−1.1	Virgo
1999 Apr 24	1997 May 1	87	16.2	−1.5	Virgo
2001 Jun 13	1997 Jun 21	67	20.8	−2.1	Sagittarius
2003 Aug 28	2003 Aug 27	56	25.1	−2.7	Capricornus
2005 Nov 7	2005 Oct 30	69	20.2	−2.1	Aries

lie in the South, while the Tharsis bulge, containing the giant volcanoes, straddles the equator).

The Martian 'day' or sol amounts (in terrestrial terms) to 24 hours 37 minutes 22.6 seconds. Therefore, the hemisphere facing us at any particular moment will again face us on the next night, 37 minutes later. Any special feature can be observed at some time during a night for a week or two in succession, after which it will be turned in our direction only during daylight; a return to original conditions takes about five and a half weeks. This can be infuriating for the observer who is trying to follow an interesting event, such as a dust storm, but nothing much can be done about it. Though observations can be made at dawn and in twilight conditions, it is really hopeless to attempt to study Mars when the Sun is above the horizon. Among planetary observers, only students of Venus and, to a lesser extent, Mercury have this privilege.

A central meridian point had to be chosen, and the small craterlet Airy-0 has been selected, in the dark area once called Meridiani Sinus, or the Meridian Bay, and now officially referred to as Meridiani Planum. Needless to say, Airy-0 is quite beyond the range of Earth-based telescopes; but Meridiani is not, and it is quite good enough to take it as the zero for longitude. The longitude of the central meridian of Mars for midnight daily is given in the *Handbook* of the British Astronomical Association, and in various almanacs, and from this it is a matter of simple addition to work out the longitude of the central meridian for any given time. The table giving the change in longitude is given in Table 6.2.

The main features to be seen on Mars are: the polar

Table 6.2. Change in longitude of the central meridian of Mars

Hours	° Long	Minutes	° Long	Minutes	° Long
1	14.6	10	2.4	1	0.2
2	29.2	20	4.9	2	0.5
3	43.9	30	7.3	3	0.7
4	58.5	40	9.7	4	1.0
5	73.1	50	12.2	5	1.2
6	87.7			6	1.5
7	102.3			7	1.7
8	117.0			8	1.9
9	131.6			9	2.2
10	146.2			10	2.4

Figure 6.2 Two views of the southern polar cap of Mars.

Top: 17 August 1971 00 – 01 UT; ω = 343°.5, 115-mm OG ×186
Bottom: 7 August 1988 2320 UT; ω = 319°, 115-mm OG ×186

caps, which wax and wane with the Martian seasons (two views of the southern cap are shown in Figure 6.2); the red tracts known as deserts; and the dark areas, which are permanent and which have been under observation ever since 1659, when Christiaan Huygens made the first disc drawing to show a recognisable feature, the triangular Syrtis Major. The polar caps are obvious when they are large (normally, of course, only one cap can be seen at a time), while the dark areas are well-defined except when Mars is experiencing one of its global dust-storms.

There is, however, one problem which has bedevilled observations of Mars ever since the 1870s. This is the myth of the canals.

The canals were first drawn in detail by the Italian astronomer G. V. Schiaparelli in 1877, though streaky features had been recorded earlier; Schiaparelli called them *canali* (channels). For some years, nobody else could see them, but in 1885 they were 'confirmed' by Perrotin and Thollon, using the powerful refractor at the Nice Observatory, and subsequently canals became all the rage. With observers such as Percival Lowell, the Martian disc began to resemble a spider's web; there were canals and double canals of all types, and dark patches christened oases. If these drawings had been accurate, then Mars must have been inhabited – as Lowell at least firmly believed, and continued to do so right up to the time of his death in 1916. The canals were even 'seen' by observers using telescopes as small as 3 in refractors. Thanks to spacecraft, we now know that they do not exist in any form; they were due to nothing more than tricks of the eye – and when straining to glimpse details at the very limits of visibility, it is only too easy to 'see' what one expects to see. Wishful thinking is the main enemy of the planetary observer, and it applies with particular force to Mars.

En passant, I once took a *Viking* map of Mars, and upon it superimposed Lowell's canal network. I half-anticipated finding that the main canals, with grandiose names such as Ganges, Gehon, Hiddekel and Eumenides–Orcus, would correspond with mountain chains, valleys or even boundaries between light and darker areas. In fact I found that there was no correlation with anything at all, so that even the great planetary observer Gérard de Vaucouleurs was wrong when, in 1956, he claimed that the canals did at least have 'a basis of reality'.

This is no place to delve further into the canal controversy; suffice it to say that it held back accurate mapping of Mars for a very long time. Another fallacy concerned the so-called 'wave of darkening'. It was once believed that the dark areas were old sea-beds filled with organic matter ('vegetation'), and that this vegetation was activated by moisture wafted from the polar regions when the ice-caps began to shrink with the arrival of warmer weather at springtime, so that there was a steady progression of darkening beginning at higher latitudes and spreading equatorwards. I have had the advantage of being able to study Mars with very large telescopes, including the Lowell refractor at Flagstaff in Arizona, and I never saw any sign of either the canals or the wave of darkening. As neither exists, I am in retrospect glad that I did not!

There is also the question of whether any of the craters recorded by the spacecraft can be seen from Earth. Certainly they are large; one, unsurprisingly named after Schiaparelli, is 500 kilometres across. It has been reported that craters were glimpsed by E. E. Barnard, using the Lick refractor in 1892, and by John Mellish in 1917, with the Yerkes refractor. Whether or not these reports are correct is debatable; none of the alleged sketches survive. In any case it is safe to say that the craters are to all intents and purposes beyond the range of Earth-based telescopes.

To recapitulate: Mars is not an easy target. Because the disc is generally small, it is essential to use a fairly high power telescope if it is hoped to see anything except the most prominent features. Of course a small telescope such as a 7.6-cm refractor or a 15-cm reflector will show something under good conditions, but for more detailed work a larger aperture is needed. A 20-cm telescope is about the minimum for a reflector; I would not personally be happy with anything below 30-cm, though opinions differ, and no doubt observers more keen-sighted than I am will disagree. Filters can be used, and some people are enthusiastic about them, but inevitably they cut down the light, and with Mars there is not too much light to spare.

Moreover, one has to wait, not only for a period around opposition, but also for the time when Mars reaches a respectable height above the horizon; and for northern observers the opposition of 2001, for example, will be very unfavourable, because Mars will be low down in Sagittarius. Conversely, the opposition of 1993 was unfavourable for southern observers, as Mars was in Gemini and was not far from aphelion at the time of opposition, so that the maximum apparent diameter never amounted to as much as 15 seconds of arc.

A disc drawing of Mars is always desirable, and in general there are enough familiar features to enable the observer to find his way around, but it is prudent to avoid looking up the longitude of the central meridian before going to the telescope, to avoid unconscious prejudice. It must be borne in mind that dust-storms and other atmospheric effects can cause dramatic changes in the planet's appearance (see Figure 6.3).

The Martian atmosphere is very tenuous. Formerly it was believed that the ground pressure was about 87 mbar, and that the main constituent was nitrogen; in fact the pressure is below 10 mbar everywhere, and the bulk of the atmosphere is made up of carbon dioxide. Nevertheless, clouds are quite common. White, isolated clouds should be carefully monitored; they change in position from night to night, and provide important information about the wind speeds – which can be rapid, though in that thin atmosphere the winds have very little force.

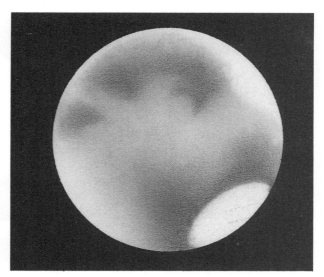

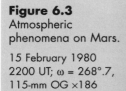

Figure 6.3
Atmospheric phenomena on Mars.

15 February 1980 2200 UT; ω = 268°.7, 115-mm OG ×186

But it is the dust storms which can be so very obtrusive. Material whipped up from the deserts can spread with remarkable speed, and there are times when Mars appears to be almost as featureless as Venus. For example, a dense dust-storm was in progress when *Mariner 9* reached Mars in 1971 – the first probe to be put into orbit round the planet. On arrival, *Mariner* could see nothing apart from the tops of the great volcanoes poking out above the dust-layer. It was only when the storm died away that the true surface features were revealed in all their glory.

Dust-storms are most frequent near perihelion, so that they tend to occur when Mars is best placed at the most favourable oppositions. Checking on their extent, distribution and degree of opacity is one of the most important programmes available to the amateur who is equipped with an adequate telescope. It is useful to make intensity estimates of the various features, on a scale from 0 (the brightness of the polar cap) through to 10 (the darkness of the night sky); in fact this is valuable whether dust-storms are in evidence or not.

One very important point concerns the phase, which is very evident except when Mars is near opposition. Indeed, the minimum phase amounts to as much as 85 per cent. If this is neglected during an observation, very serious errors will be introduced. Outlining a disc with the required amount of phase is a wearisome business, and it is certainly better to use a prepared disc. The set given here (Figure 6.4) can be photocopied and used; of

course, the phase for any given night is given in the BAA *Handbook*. Generally speaking, observers use a fixed scale of two inches to the full diameter of the planet, though this is a matter for personal preference.

Begin by taking your prepared blank, with the correct phase, and putting in the main details – that is to say the polar cap, if visible, and the visible dark areas. Mars rotates much more slowly than, for example, does Jupiter, and therefore it is permissible to take rather more time to sketch in the main features; all the same it is not wise to linger for too long. Check to ensure that the basic outline is as accurate as possible, note the time (in GMT) and do not subsequently change the outline. Then check various magnifications to see which is the highest power which will give a really sharp image; survey the planet with this power, and then put in the minor details, notably any clouds or dust-storms which

Figure 6.4
Prepared disc outlines of Mars. Each disc is drawn to show a degree of phase. From left to right, top to bottom, the phase is: 85%; 90%; 95%; 98%. If the discs are enlarged on a photocopier to 140% they will be 2 inches in diameter, as recommended in the text.

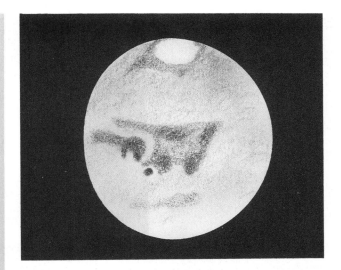

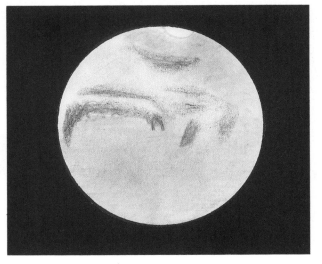

Figure 6.5 Two views of Mars, prepared by the author.

Top: 28 August 1973 0235 UT; 27-inch OG (Johannesburg) ×580
Bottom: 2 September 1973 2345 UT; 12½-inch reflector ×360, ω = 003; seeing = II

can be made out. When you have finished, again note the time, and add your name, the aperture and type of telescope used, longitude of the central meridian, and the seeing on the usual Antoniadi scale, which goes from 1 (perfect; seldom attained) down to 5 (so poor the sketch would not normally be attempted). If any of the detail is omitted, the observation promptly loses most of its value.

Despite the obvious difficulties, observers should begin work as long as possible before opposition, and continue for as long as possible after it – which amounts to months for an observer who has the use of, say, a 30-

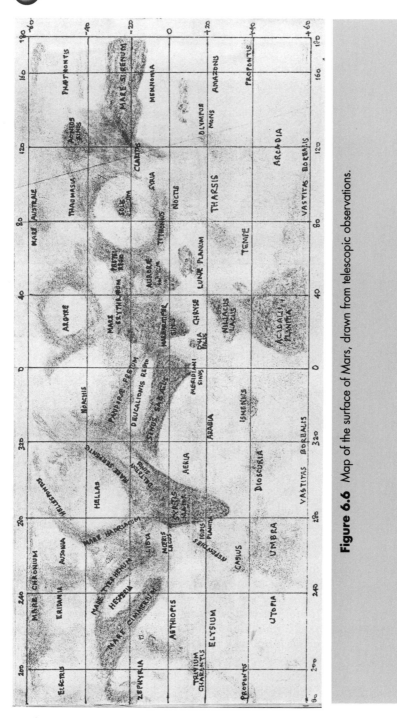

Figure 6.6 Map of the surface of Mars, drawn from telescopic observations.

or 40-cm telescope. Artistic ability is always a help, but even those with no such skill (such as myself!) need not be disheartened – see Figure 6.5; the main essential is accuracy.

The map given here (Figure 6.6) has been compiled from observations made with telescopes up to 39 cm in aperture. It is oriented with South at the top. This has long been the usual convention, because all normal astronomical telescopes give an inverted image, and until relatively recently most observers were based in the northern hemisphere. Today there is an increasing tendency to put North at the top; it does not matter in the least, but it is only sensible to be consistent about it.

The original nomenclature, in which features were named after planetary observers, was abandoned by Schiaparelli in 1877; he allotted new names, and these were accepted and extended by E. M. Antoniadi, who was arguably the best of all pre-Space Age planetary observers. It is essentially Antoniadi's system which we use today, though the space results have made revisions necessary. For example the old Nix Olympica, or Olympic Snow, has been re-named Olympus Mons, or Mount Olympus; it is a towering volcano, three times the height of Everest, rather than a snowy patch. Mare Acidalium has become Acidalia Planitia; it is not a sea – there are no seas on Mars. Noctis Lacus is not a lake, but a system of canyons beside which our Grand Canyon in Colorado would seem very puny, so that it has become Noctis Labyrinthus. On the map I have naturally followed the official names, though old-fashioned observers such as myself still instinctively think of Mare Sirenum, Mare Tyrrhenum and Sinus Sabeus!*

Let us now begin a brief survey of the accessible features on the Martian disc.

The Polar Caps

These can be among the most striking features on the planet. At maximum a cap may extend down into tem-

---

* *Mons*: mountain or volcano. *Planitia*: smooth, low-lying plain. *Planum*: plateau; smooth, high area. *Terrae*: lands, names of ten given to classical albedo features – see below. *Labyrinthus*: valley complex; network of linear depressions. *Vastitas*: extensive plain.

Table 6.3. Martian calendar

	Days	Sols
S. spring (N. autumn)	146	142
S. summer (N. winter)	160	156
S. autumn (N. spring)	199	194
S. winter (N. summer)	182	177
	687	669

perate latitudes; at minimum it may almost or quite vanish. The caps are not identical, because conditions of temperature in the southern hemisphere are more extreme than in the North; the southern summers are shorter and hotter, the southern winters longer and colder (a Martian calendar is given in Table 6.3). This means that the southern cap can become more extensive than the northern cap may ever be. Moreover, the caps differ in structure. In each case there is a residual cap of water ice overlaid by a coating of carbon dioxide ice, but in the summer the carbon dioxide layer on the North cap disappears, which does not happen in the South.

Draw the outline of the cap very carefully – not too easy when, as often happens, the boundary is indistinct. Patches of whiteness may be left as the cap retreats, and these should be recorded with particular care because they provide information about Martian relief. Interesting irregularities in the border can be found, which change markedly from night to night. If a cap is on view, check for indications of whiteness at the opposite pole, even though the cap there will presumably be turned away from us (the tilt can be checked from the BAA *Handbook* or an almanac; for example, in December 1994 the tilt amounted to +22°, indicating that it was the North pole which was presented to us). When a cap is shrinking, a dark 'collar' can usually be seen along its border, and can hardly be overlooked, even though we now know that the older observers were wrong in attributing it to marshy ground or even a temporary system of lakes.

The Great Diaphragm

The band of dark markings running round the planet, lying mainly South of the equator but in some places

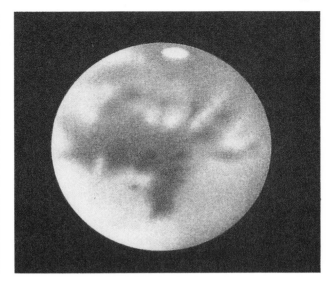

Figure 6.7 Syrtis Major, the darkest feature on Mars.

20 October 1988
1830 UT; ω = 303°.8:
δ = 21 inch:
φ = −23°.5,
115-mm OG ×186;
exceptional seeing and transparency

crossing it, has often been nicknamed the Great Diaphragm. The most prominent feature is *Syrtis Major*. In the North, the main dark area is *Acidalia Planitia* (the former Mare Acidalium). It does not take long to learn the main characteristics of the Martian surface, though major dust-storms can alter the entire aspect and confuse even the experienced observer.

Syrtis Major

This triangular feature, formerly known as the Hourglass Sea or Kaiser Sea, is always dominant. It is the darkest feature on Mars, and is well-defined, as shown in Figure 6.7. When it is on the disc, and is not veiled by dust, any small telescope will show it when Mars is near opposition.

Hellas

Once regarded as a snow-covered plateau; now known to be the deepest basin on Mars. It measures 2200 × 1800 km, and is about 3 km deep, reckoned from the level on Mars where the atmospheric pressure is 6.1 mbar. In brightness it is very variable. At times it can be so brilliant that unwary observers have mistaken it for a polar cap. It is bounded by darker regions, *Serpentis Terra*, *Hadriacum Terra* and *Hellespontus*.

Aeria

This is a relatively featureless, ochre tract adjoining Syrtis Major and *Sabæa Terra*. The adjoining *Arabia* is similar.

Dioscuria

This is an area of rather elusive, dark shading. Near it is the rather more obvious *Ismenius*.

Sabæa Terra

Sabæa Terra (formerly Sinus Sabæus) is a very prominent, dark feature, adjoining Syrtis Major but often appearing slightly separated from it. It extends to the prime meridian of Mars, ending at the *Meridiani planum* and the small, 'pronged' feature known as *Fastigium Aryn* and nicknamed Dawes' Forked Bay (after a nineteenth-century British astronomer, the Revd W. R. Dawes). Fastigium Aryn can be seen with a 20-cm telescope under good conditions, but it is a delicate feature.

Pandoræ Terra

Formerly known as Pandoræ Fretum, this is another elongated, dark area, south of Sabæa, from which it is separated by the lighter region of Deucalionis Regio.

Noachis

A light area south of Pandoræ, Noachis is distinct from Hellas to the one side and Argyre to the other, but is not nearly so well-defined.

Margaritifer Terra

Triangular, its shape is not unlike that of Syrtis Major, but Margaritifer (once known as Margaritifer Sinus) is much smaller and less dark. Close to it is the small, dark patch of *Oxia*, formerly known as Oxia Palus.

Erythræum Plana

This forms a darkish, but sometimes ill-defined, area between Margaritifer and Argyre. Formerly known as Mare Erythæum, it merges into both Margaritifer and Auroræ Planitia.

Argyre

A deep basin, 740 km in diameter, Argyre is the most prominent feature of its type on Mars, with the exception of Hellas. It too can be very bright, and dust-storms very often start in or near it.

Australe Planum

Once known as Mare Australe, this dark region is too far south to be well seen.

Chryse Planitia

Chryse Planitia is the 'Golden Plain', north of Margaritifer. It is quite distinct, and it was here that the lander of the *Viking 1* spacecraft came down on 20 July 1976 (lat. 22° 4′ N, long. 47° 5′).

Acidalia Planitia

Apart from Syrtis Major, the former Mare Acidalium is the most prominent dark area on Mars; it is wedge-shaped, and has a southward extension (*Niliacus*). Acidalia often seems to merge with the dark band bordering the retreating northern polar cap. Adjoining it is the ochre tract of *Tempe*.

Auroræ Planitia

Yet another somewhat triangular feature in the Great Diaphragm, Auroræ Planitia (formerly Auroræ Sinus) is not unlike Margaritifer, but is less well defined. North of it is the darkish, small *Lunæ Planum* (formerly Lunæ Lacus), which can be quite conspicuous.

Solis Planum

This is a notoriously variable feature, both in size and shape as well as in brightness. It has been nicknamed 'the Eye of Mars' and was once called Solis Lacus, 'the Lake of the Sun'. It was the first choice as the zero for longitudes, and is surrounded by a light area which is bounded by the darker *Tithonius*, *Protei*, and *Thaumasia*.

Noctis Labyrinthus

This can often be seen as a somewhat darkish area, and was originally thought to be a lake (Noctis Lacus) rather than a system of canyons. It is never prominent, but is worth finding.

Tharsis

Tharsis is the great 'bulge' holding the giant volcanoes of Pavonis Mons, Ascræus Mons and Arsia Mons, with Olympus Mons in the same area. These volcanoes can be seen as small patches under good conditions. Well to the south of Tharsis is the ochre tract of *Arcadia*.

Sirenum Terra

A very conspicuous, dark feature in the Great Diaphragm, which was once called Mare Sirenum. South of Sirenum is the fairly well marked *Aonius Sinus*, and the less obvious *Claritas* lies between it and the dark features surrounding the Solis Planum area.

Phæthontis

A light region south of Sirenum. The Russian probe, *Mars-3*, landed here in December 1971.

Memnonia

Memnonia and the adjoining tract, *Amazonis*, lie north of Sirenum. At latitude 40° N the dark streak of *Propontis* is fairly evident.

Electris

A bright tract in the far south. Still further south is the dark *Chronium*, always very foreshortened. Another bright area here is *Eridania*, where the Russian probe, *Mars-2*, landed in November 1971.

Cimmeria Terra

Cimmeria Terra and *Tyrrhenna Terra* (formerly Mare Cimmerium and Mare Tyrrhenum) make up a conspicuous part of the Great Diaphragm; they are separated by the well-defined light region of *Hesperia*.

Ethiopis

An ochre region to the north of Cimmeria. Still further north is *Elysium*, now known to be one of the two major volcanic regions of Mars (*Tharsis* is the other).

Trivium Charontis

This is a small, but sometimes distinct, dark patch in the Elysium area. On the 'canal period' maps it was shown as the centre of a whole system of streaks.

Ausonia

Ausonia is yet another fairly well defined, light region, between Eridania and the Hellespontus.

Libya

A small, but fairly dark region linking Tyrrhena with Syrtis Major. There is also a 'bay' here, Mœris Lacus on the old maps, which appears as an indentation in the boundary of Syrtis Major.

Umbra

Umbra is a dark area in the far north. Adjoining it is *Utopia*, the bright plain where the lander of *Viking 2* came down on 3 September 1976. South of Umbra is the 'Wedge of Casius'. A streak running from here to *Isidis Planitia*, bordering Syrtis Major, is still known commonly as Nepenthes, one of the few 'canal period' names retained, though Nepenthes looks nothing like a channel.

Vastitas Borealis

A darkish area in the far north, running right around Mars and often covered by the polar deposit.

I have said nothing about photography here because, frankly, good photographs of Mars require highly sophisticated equipment. CCDs are coming into use, but it is probably true that, for sheer scientific value, amateur observations are more useful than amateur photographs.

The satellites, Phobos and Deimos, are always elusive, and the best way to glimpse them is to cover Mars itself with a suitable occulting disc. Both satellites can be seen with a 30-cm telescope when conditions are good, but not easily.

All in all there is much to be done, and the Red Planet remains probably the most intriguing of all the Sun's family – even if there are no canals, and the brilliant-brained Martians of Lowell and his colleagues have long since been relegated to the realms of science fiction.

Bibliography

Baker V R, *The Channels of Mars*, Hilger, 1982   ***Books***
Beatty J K and Chaiken A (Eds), *The New Solar System*, Cambridge, 1990
Carr N H, *The Surface of Mars*, Yale, 1982
Cattermole P, *Planetary Volcanism*, Ellis Horwood, 1989
Cattermole P, *Mars*, Chapman & Hall, 1994
Moore P, *The Amateur Astronomer*, Cambridge, 1992
Moore P, *New Guide to the Planets*, Sidgwick & Jackson, 1994
Sheehan W, *Planets and Perception*, Arizona Press, 1988

Periodicals

Sky & Telescope (Cambridge, Mass.): *monthly*
Practical Astronomy (Concept Publications): *monthly*
Astronomy Now (Intra Press): *monthly*
Handbook of the British Astronomical Association:
    annual
The Yearbook of Astronomy (Macdonald): *annual*

Chapter 7
Jupiter

Terry Moseley

Jupiter is undoubtedly the most rewarding single object in the sky for the amateur observer. Of all the planets it has by far the largest mean apparent diameter. Its maximum apparent diameter is exceeded only by that of Venus, but only when the latter is a hardly observable slender crescent; and little detail is observable on Venus at the best of times. Even its minimum observable disc exceeds the maximum observable discs of Venus, Mars and Saturn combined.

Further, its disc presents more detail and more variation than any other planet, and the incredible extravaganza that is the atmosphere, as revealed by spacecraft, is almost tantalisingly within the reach of a good amateur telescope in moments of excellent seeing. But the pictures of the surface revealed by the *Pioneers* and *Voyagers* were transitory, and no probe has yet revealed what lies beneath the fantastic, multicoloured visible surface.

Given that it is larger and more massive than all the other planets combined, and radiates more energy than it receives from the Sun, that it has the most intense magnetic field of any planet, and that it has the most observable family of satellites, it is little wonder that it receives intense study by both amateurs and professionals. Since Jupiter is so bright, it can be observed in twilight less than three weeks from conjunction, especially with a refractor, and so can be followed for over ten months each year from somewhere or other on Earth.

The most important physical data about Jupiter are given in Table 7.1. The most interesting aspect for the observer is the very rapid rotation period of under 10

Table 7.1. Jupiter: basic physical data

Mean synodic period	398.88 days
Mean sidereal period	4332.59 days
	(11.862 years)
Mean distance from Sun	5.20280 AU
Equatorial diameter	142,800 km
Polar diameter	134,200 km
Oblateness	1:15.4
Inclination of orbit to ecliptic	1.303°
Orbital eccentricity	0.04849
Sidereal axial rotation	9.842 hours
Inclination of equator to ecliptic	3.12°
Reciprocal mass (Sun = 1)	1047.355
Mass (kg)	1.899×10^{27}
Mean density (water = 1)	1.32
Escape velocity	59.6 km/sec
Surface gravity (Earth = 1)	2.69
Volume (Earth = 1)	1323
Mean visual opposition magnitude	−2.6
Maximum visual opposition magnitude	−2.9
Albedo	0.52

hours. This means that features on the disc are presented to the observer in fairly quick succession, and it is possible to record the appearance of the whole disc in a single night. For example, I was lucky enough to be able to observe six complete rotations during the 1966–67 opposition!

The fast rotation facilitates one of the most valuable observational techniques: the timing of the transits of features across the central meridian (CM), the line joining the north and south poles of the planet. The rapid rotation causes noticeable flattening of the disc, to a ratio of 1:15.4, caused by centrifugal force. It also makes the main belts and zones follow a generally east–west orientation, making it quite easy to orient the disc to determine the equator and the north–south line, either in the eyepiece or on a photograph or CCD image.

Instruments

One of the greatest Jupiter observers, Stanley Williams, used only a 6-inch reflector, but most serious students of the planet now would look for at least an 8-inch, although a good 5-inch OG can reveal surprising detail.

This is not the place to debate the relative performance of refractors and reflectors, but good resolution, high contrast and faithful colour rendition are essential. A good, long focal ratio Newtonian reflector, a Maksutov, or an apochromatic refractor is probably the best but, as in every field, the quality of the observer is the most important factor, and good results can be obtained with any reasonable instrument.

Large apertures can be more adversely affected by bad seeing, and anyone with a large reflector – say 16 in or more – should make the largest possible off-centre mask (i.e. diameter = primary radius – secondary radius) to use in bad seeing.

Eyepieces should have high contrast and definition, rather than a wide field: orthoscopics are ideal, but any type of good quality will suffice. The ideal magnification will depend on both instrument and seeing – anything from about ×120 upwards; in exceptional conditions more than ×600 can be used in a good, large telescope.

An equatorial mount with a drive, or at least slow motions, is preferable, but a good Dobsonian or other altazimuth is not to be despised. Modern computer-controlled 'smart drives' on some Schmidt–Cassegrain telescopes also make observing very easy.

Jupiter's Disc

Appearance of the Disc

Even a small telescope reveals the main dark belts and lighter zones crossing the disc. The Equatorial Zone (EZ) is bounded by the North and South Equatorial Belts (NEB and SEB), and other belts and zones of varying prominence and permanence alternate towards higher latitudes until they fade into the general obscurity of the polar hoods. These dusky hoods are partly a genuine feature of the planet itself, as revealed by spacecraft, and partly the result of looking at the surface at an increasingly oblique angle at higher latitudes. The standard nomenclature for the Jovian belts and zones is given in Figure 7.1.

Sometimes a belt may appear to be double; if so, the components are labelled North and South, e.g. SEB(N) and SEB(S). In such cases, the area in between is referred to as a zone – in this case the SEBZ. Sometimes

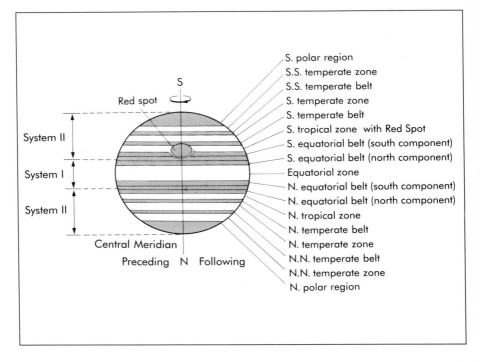

S. polar region
S.S. temperate zone
S.S. temperate belt
S. temperate zone
S. temperate belt
S. tropical zone with Red Spot
S. equatorial belt (south component)
S. equatorial belt (north component)
Equatorial zone
N. equatorial belt (south component)
N. equatorial belt (north component)
N. tropical zone
N. temperate belt
N. temperate zone
N.N. temperate belt
N.N. temperate zone
N. polar region

S

Red spot

System II

System I

System II

Central Meridian

Preceding N Following

Figure 7.1 The belts and zones of Jupiter.

a zone has a narrow, dark belt in it, referred to as a band; often one is seen in the middle of the EZ. And occasionally a belt or component may be missing altogether, being replaced by white 'zone' material.

As Jupiter's orbit is inclined by only 1° 18′ 20″ to the ecliptic, and as its own equator is tilted only 3.07° to its orbit, the polar regions are never well presented towards the Earth. The low inclination of the axis means that seasonal effects are minimal, and in fact the polar regions never receive much radiation from the Sun, which may explain their relative quiescence. The lowish orbital eccentricity of 0.048 gives relatively little variation in distance from the Sun.

The limb, or apparent edge of the disc, is perceptibly darker than the centre of the latitudes, because of the lower angle of sight and solar illumination, and any features near the limb will be less obvious, due to this darkening and foreshortening. Thus, at the edge of the disc, the belts become more obscure, and the bright zones are noticeably darker, an effect easily seen during the satellite transits, or on photographs.

The Atmosphere

Jupiter's atmosphere is comprised mainly of hydrogen and helium, with lesser amounts of ammonia and simple hydrocarbons. The visible clouds seem to be formed mainly of ammonia crystals, ammonium hydrosulphide and ice. The highest and coldest regions are the whitish zones, and orange features such as the Great Red Spot. The lowest and warmest areas are the brownish belts, and deepest of all are the dark projections on the southern edge of the NEB, which often appear bluish-grey. Jupiter gives out about twice as much heat as it receives from the Sun, so the interior must be very hot.

Rotation: Systems I and II

The visible surface of Jupiter is the outer layer of the atmosphere, and different parts of this rotate at different, and variable, rates. There are, however, two fairly distinct rotation zones: the equatorial region, which comprises the region from the southern component or half of the NEB to the northern component or half of the SEB, has a defined mean rotation period of 9h 50m 30.003s, and is called System I. All the rest of the planet has a mean assigned period of 9h 55m 40.632s, and is called System II. (The presumed rotation of the inner core, 9h 55m 29.711s, as determined from radio observations, is called System III, but this need not concern us here.) Actual rotation speeds vary considerably from these means, but features are referred to either one system or another, depending on location. The two systems are commonly abbreviated, either to λ^1 and λ^2, or to ω^1 and ω^2.

Some of the most interesting phenomena are the 'jet streams', first discovered by amateurs, consisting of groups of spots moving very rapidly along the edges of some of the belts. The relative speed of these groups can be up to several degrees of longitude per day. They move in a preceding direction (decreasing longitude) at the equatorial edges of each belt where they occur, and in a following (increasing longitude) direction on the poleward edges of the belts. Space-probe observations indicate that these are at least semi-permanent, but only the major ones are observable from Earth.

As seen in an inverting telescope, the planet rotates from the 'following' to the 'preceding' side of the field, and these terms denote the equivalent edges of the disc. Allowance must be made for any optical system producing a reversed or mirror image, and in fact these should be avoided if possible.

Nomenclature of Features

More or less standard terms have developed for the various types of features seen on Jupiter. 'Spots' are round or oval, and can be bright or dark. If elongated they are referred to as 'streaks' (bright), or 'rods', 'bars' or 'barges' if dark. Very small, dark spots are sometimes termed 'condensations', particularly if in a belt. Their white counterparts are often called 'nodules'. 'Ovals' are larger than spots, and usually white, though sometimes they are grey. 'Festoons' or 'plumes' are usually thin and faint, and sometimes form a bridge, usually oblique and curving, across a zone. Sometimes they curve back to join further along the same belt, in which case they may be known as 'garlands'. These are often found joining dark 'projections' or 'humps' on the southern edge of the NEB. 'Rifts' are bridges of light material crossing a belt from one zone to another. A 'knot' is a thicker or darker segment within a thin belt. 'Notches' are small versions of 'indentations' in the edge of a belt.

Colours of the Disc

The overall appearance of the disc is creamy white, crossed by the dark bands, which are predominantly grey-brown, but other fainter shades of red or purple can also be seen. The vivid colours of the spacecraft photos are not seen by the visual observer, partly at least because the photo colours were greatly enhanced. Very dark markings, such as some projections on the southern edge of the NEB, may appear to have a bluish tint.

Of course, observations of colour can only reliably be made with a reflector, or a high quality apochromat, with a good quality eyepiece, and when Jupiter is high in the sky to minimise the refractive effects of the Earth's atmosphere.

The use of colour filters, such as Wratten No. 80A or No. 82A (light blue) will enhance the contrast between the belts and zones, while some dark markings, such as festoons, are enhanced by a Wratten No. 12 (yellow) or No. 21 (orange) filter, indicating a bluish tinge.

Photographic or CCD images taken through colour filters will also show up the colour differences of various features.

The Great Red Spot

The longest-lived feature on Jupiter, the Great Red Spot (GRS), was first definitely recorded by Cassini in 1665, and perhaps by Hooke in 1664. It then became especially noticeable in 1878, and has been observed almost continuously, though sometimes it is very faint, ever since. As it has drifted in longitude by about 1200 degrees in the last 100 years, it obviously cannot be a solid surface feature. It appears to be a meteorological phenomenon – a sort of gigantic hurricane or atmospheric whirlpool, and there are indications that it may be decreasing in size, and so may not be permanent. Even when very faint or colourless, its position is always indicated by the Red Spot Hollow (RSH), an indentation in the southern edge of the SEB.

The spot was obvious in the 1960s and early 1970s, but recently it has varied in prominence. Interestingly, it is not the variations in size which affect visibility as much as the changes in colour and contrast with the background. The colour has been described as varying between brick red, through orange, to a faint pink or dusky or yellowish grey. When faint or almost colourless its visibility can be enhanced by blue, green or magenta filters (Wratten Nos 80A, 82A, 58 or 30).

The GRS drifts in both latitude and longitude, though these variations do not seem to be correlated with other aspects such as its prominence. It usually lies around latitude 22°S.

The GRS rotates anti-clockwise in about 12 days or less, although the centre seems to have a period of less than 9 days. Material wells up from the interior in the middle, and spirals out to the edge. This uprising makes the centre some 80 km higher than the surrounding clouds, and therefore colder. Sometimes neighbouring spots get sucked in to the Red Spot, and altogether it is an amazing area of the planet.

Recent theories indicate that the colour may be due to crystals of red phosphorus created by phosphine (hydrogen phosphide) reacting with ultra-violet radiation from the Sun.

Currently the GRS is about longitude 45° (System II), and is reasonably prominent.

The South Tropical Disturbances (STDs)

These are another example of fairly long-lived Jovian phenomena, and were first recorded by amateurs. An STD consists of a grey patch or barrier across the STZ, the greatest of which lasted from 1901 to 1939. It first appeared as a small, dark spot, and gradually expanded in longitude. It produced an amazing circulating current, whereby spots retrograding on the SEB(S) jet stream were curved around the concave preceding end of the STD to the STB(N) jet stream going back in the opposite direction. There have been at least seven shorter versions of the STD phenomenon since 1939, most of which started at the preceding end of the GRS. The last STD appeared in 1993, and is still present at the time of writing.

SEB Revivals

Occasionally the SEB(S) component disappears, and this is usually accompanied by increased prominence of the GRS. Then a small, dark streak appears in the latitude of the SEB(S), and many small, dark and white spots pour out from this area, some of which retrograde on the SEB(S) jet stream, the others moving the opposite way on the SEB(N) jet stream. A scene of chaotic activity ensues, after which the GRS fades, and the SEB revives. This can occur at intervals of between three and thirty years, and last happened when the SEB faded in the summer of 1992, followed by a revival in April 1993, which is still ongoing.

STB/STZ White Ovals

Three bright white oval spots appeared around 1940, lying in the STZ. They started as three longer, brighter sectors of that zone, and gradually contracted to become fairly prominent white spots, lying in indentations in the southern edge of the STB. They are gradually moving into the STB and disappearing. They are called FA, BC and DE, after the ends of the bright sectors from which they formed. They regularly overtake the GRS, forming nice conjunctions.

Observation

Visual Observing

Disc Drawings

Most people will start their Jupiter observing by drawing the disc (see Figure 7.2). First, draw the main belts and zones, recording their latitudes, widths and intensities as well as you can. Then quickly sketch in any major features such as spots, streaks or projections, and note the time to the nearest minute. Next, add the fine details, relative to the features already drawn, starting with those near the preceding edge, which disappear first due to the rapid rotation. Then add shadings, and note any colour and brightness intensities on a scale of 0 (brightest, equal to the very brightest spots) to 10 (black). Often no features will be as bright as '0', and '10' is very rare. These can then be incorporated into a final drawing made away from the telescope. Another full disc drawing should be made every two to two-and-a-half hours. Unless you are an experienced astronomi-

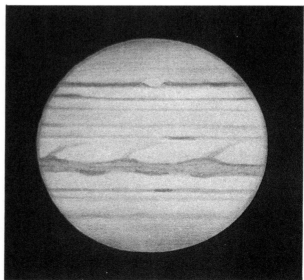

Figure 7.2 A drawing of Jupiter's disc, one of the first tasks in visual observation.

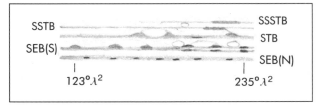

SSTB

SEB(S)

$123° \lambda^2$

SSSTB

STB

SEB(N)

$235° \lambda^2$

Figure 7.3 A strip sketch recording features seen in Jupiter's southern belts.

cal artist, it is probably best not to attempt to draw in colour at the eyepiece: your faint light will not show colours properly.

Strip Sketches

Detailed drawings can be made of a particular area, usually of a particular belt and zone. They can be combined with transits (see below) and added to as the rotation brings new areas into view. A strip sketch is shown in Figure 7.3.

Transits

Once familiar with the general appearance and rotation of the disc, the observer should record the times at which visible features cross the central meridian (CM), the line joining the north and south poles, to the nearest minute. With practice, good seeing, and reasonably powerful optics (×200+), an accuracy of one to two minutes can be obtained, representing a longitudinal measurement of about 1°. This is more than enough to show the changes in rotation speed and the relative drifts of different features, which are among the most fascinating and intriguing of Jovian phenomena.

For features extending in longitude, e.g. dark streaks or white ovals, record the transit times of preceding and following ends, as well as of the centre. This will show any changes in size, and whether a feature is growing or shrinking at one end or both.

Each feature should be described according to type, shape, colour, intensity and location, along with its transit time; e.g. '22.19: p. end of large elongated white oval in NTZ'. It should also preferably be identified with a code letter or number on an accompanying disc or strip sketch, particularly if there is a lot of detail. The data should later be entered into a log book, with a serial number: a suggested layout is shown in Table 7.2.

Table 7.2. Suggested layout for Jupiter transits log book

Date: 25/26 Feb 1994. 200-mm reflector, ×220

Time (UT)	No.	Description	λ^1	λ^2	Seeing	Notes
22.15	38	Small bright spot, STZ	–	31.7	fair	
22.21	39	p. edge, dark rod in NTB	–	35.3	fair	fading?
22.22	40	Centre of bay, NEB(S)	202.1	–	good	

If there are interesting features just past the meridian when you start observing, or after interruption by cloud, particularly if you have not observed for a while, it is worth estimating their transit times. These times should be noted as estimates only, with an assessment of likely accuracy. An experienced observer can do this with sufficient accuracy to provide confirmation of the appearance (or continued existence) or form of interesting features, if not exact longitudes. The same can be done for features not yet at the CM if observing is about to be interrupted.

Reduction and Analysis

The *Handbook* of the British Astronomical Association (BAA) gives the longitude of the CM for both Systems I and II for 0 hours UT each day, yearly, together with tables giving the changes in longitude over time. From these, the longitude of any feature can quickly be calculated according to the appropriate System. Plotting the derived longitudes on a graph on the vertical axis (see Figure 7.4) will show the relative drifts of all the features you have recorded several times or more, as long as the gaps in observing have not been too long.

Unless you are blessed with excellent weather and seeing, a good telescope, and plenty of time, you are unlikely to be able to record features often enough to provide useful results on your own. It therefore makes sense to join an organisation such as the British Astronomical Association or the Association of Lunar and Planetary Observers (ALPO), so that your results may be combined with those of other observers to give a much fuller picture.

Still, it is very satisfying to get a relatively quick analysis of what you have seen, and there is no harm in doing this as long as you do not rely too much on your own results.

Latitude Measurement

Measurement of the latitudes and widths of belts, zones, and other features can be done from high quality CCD images or photographs. If you use one of these techniques, obtain some images with the limb over-exposed, so that the edge is not lost due to limb-darkening.

Otherwise, latitude measurements require a graticule eyepiece or a micrometer. The former is cheaper, but not so accurate, partly because it is often difficult to ascertain the exact position of the equator; occasionally there is an Equatorial Band, but it sometimes lies south of the true equator. Nevertheless, such measures are better than nothing.

Align the horizontal axis absolutely parallel to the belts and zones. Then measure the positions of the poles along the graticule scale, making sure that the image is bright enough to show the very edge of the disc. The edges and centres of belts, zones and other features such as the GRS can then be assigned fractions of the dis-

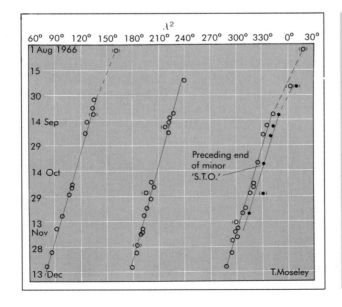

Figure 7.4 Plotting derived longitudes to demonstrate relative drift of features.

tance from the equator (0°) to the poles (90°). Half-way would give a latitude of 45°, a quarter of the distance from the equator would be 22.5°, etc. Interpolation between the graticule marks will usually be necessary but, with practice, reasonable accuracy can be achieved.

Filar micrometers vary in design and quality, but none are cheap. With careful use on a good, driven telescope, they can give very accurate results, and will probably only be superseded by measurement of the increasing number of excellent CCD images now being obtained.

The method of use depends on the model, with which instructions should be provided, but general guidelines include making several, preferably four or five, measurements, moving the wire in alternate directions. If using an ordinary achromatic refractor, it may be difficult to measure the precise edges of belts due to a false colour.

Observations of the Satellites

Jupiter has 16 moons, of which only four are observable by amateurs. Known as the Galilean moons, after their discoverer (although Marius may also have seen them at about the same time), they are Io, Europa, Ganymede and Callisto, often simply designated I, II, III and IV. Ganymede is the largest moon in the Solar System, and is even larger than Mercury. The diameters, distances, mean opposition magnitudes and revolution periods are given in Table 7.3.

Some people claim to be able to see them with the naked eye; indeed, were it not for their proximity to Jupiter, all would be visible without optical aid. Even binoculars, or the smallest telescope, will show them when they are furthest from Jupiter, but a larger instrument is needed to show their eclipses, and probably at least a good 5-inch refractor or 6-inch reflector to show the occultations and transits properly.

Table 7.3. Jupiter's satellites: physical data

Moon	Diameter (km)	Magnitude	Mean distance (km)	Mean sidereal period (days)
Io	3659	5.0	421,600	1.769
Europa	3138	5.3	670,900	3.551
Ganymede	5262	4.6	1,070,000	7.155
Callisto	4800	5.6	1,880,000	16.889

Commensurability

There is an interesting relationship in the motions of satellites Io, Europa and Ganymede. All three can never undergo the same phenomenon at that same time. Thus, two may appear in transit together, but if so, the third will be in eclipse or occultation. On rare occasions, all may appear to be 'invisible' simultaneously, being either in transit across the disc of Jupiter, or in eclipse or occultation behind it. This is next predicted to occur on 1997 August 27, from 21.37 to 21.52, on 2001 November 8, from 16.26 to 16.42, and on 2008 May 22, from 03.50 to 04.08.

Satellite Phenomena

All the moons have very low orbital inclinations, so they regularly appear in transit across the disc, or are occulted or eclipsed by it each revolution. Only when Jupiter's own axis is tilted near its maximum towards Earth does the outermost satellite, Callisto, appear to pass above or below the planet's disc. This was the situation in 1994.

It is worth timing the start and end of each transit and occultation, noting the time of first contact (edge of moon touching edge of planet), mid event (planet's limb bisecting moon) and second contact (commencement of total immersion). Do the same for the end of the event (third and fourth contacts). Similar observations can be performed for the ingress and egress of the shadows of the moons, which precede them across the disc before opposition, and follow them after it.

The shadows can be used to compare the darkness of other features on Jupiter, which will often appear very dark, until the inky blackness of a satellite shadow appears for comparison. The moons themselves can often appear very dark when in transit across a bright zone, but this is merely contrast, as can be seen by comparison with a shadow. The shadows always appear somewhat larger than their corresponding moons, especially in the cases of Ganymede and Callisto, indicating that the shadow has a large penumbral component.

As a moon starts its transit it will appear bright against the planet's dark limb, but will soon lose contrast, and may even disappear in instruments of moderate magnification, before appearing dark against the brighter centre of the disc. Europa, however, usually remains bright throughout transit because of its higher albedo.

The easiest satellite events to observe are the eclipses,

especially when Jupiter is near quadrature, when they occur at their greatest apparent distances from the disc. At such times, even a 60-mm OG will show them. It is useful to time the beginnings and endings of eclipses and the reappearances from eclipses. Timing the last moment of visibility as an eclipse commences is simple, as is the first sighting of reappearance, if you know when and where to look. Both events depend on telescope performance and conditions: a large telescope will show a moon which is going into eclipse, after it has disappeared in a 60-mm refractor. Timing eclipses is akin to variable star observing: compare the relevant moon with its neighbours until you see it either start to fade, or return to normal brightness.

Only Ganymede and Callisto, and very rarely Europa, ever appear far enough from the planet for the whole of an eclipse to be seen. After opposition, the eclipse has usually already started before the satellite reappears from occultation, so that only the reappearance can be seen.

Before opposition, the satellite often goes into occultation behind the planet before the eclipse ends. The closer to the date of opposition, the nearer the eclipse events will occur to Jupiter. Conversely, near quadrature, they will occur at their furthest from the disc.

Mutual Satellite Phenomena

Because their orbits are aligned almost towards the Sun and the Earth, the satellites periodically occult and eclipse each other, and such events can reveal quite a lot about the surface features and brightness of the moons (see Figure 7.5, a and b). The last set of mutual events was in the period 1990–1992, and another series will commence in 1996. A moderate telescope is needed to observe the occultations properly, but a more modest instrument is usually sufficient to show the effect of the eclipses. Those events which occur while a satellite is in transit are very rare, but spectacular to observe. Accurate photometry of mutual eclipses, or even occultations, can provide very useful data.

Combined Observations

At times when a spacecraft is approaching or orbiting Jupiter, it is very important to have up-to-date informa-

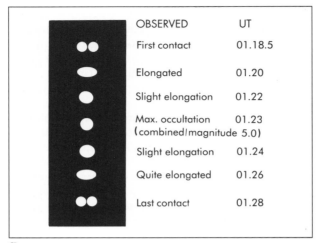

OBSERVED	UT
First contact	01.18.5
Elongated	01.20
Slight elongation	01.22
Max. occultation (combined/magnitude 5.0)	01.23
Slight elongation	01.24
Quite elongated	01.26
Last contact	01.28

a

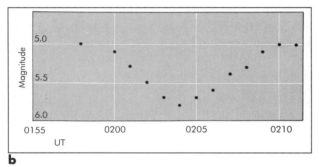

b

Figure 7.5
Jupiter's satellites in occultation.
a Plot over time as two satellites appear to merge and separate: *II occults I* (Pred. $01^h 19^m.1 - 01^h 27^m.7$. Predicted drop in light = 39%).
b Light curve showing the eclipse of Io by Europa, 2 February 1991. $4\frac{1}{4}$-inch OG ×240. seeing = III (Ant)

tion on what is happening on the planet, and the instruments and talents of amateurs and professionals are often combined in a coordinated observing programme. Any experienced observer with a good telescope can play a very useful part in such programmes. Details will be available in connection with the Galileo mission, for example, from organisations such as the BAA, and the Association of Lunar and Planetary Observers (ALPO) in North America.

Photographic and CCD Observation

Jupiter is the most rewarding of the planets for this work, because of its large apparent disc and its high surface brightness. It is likely that CCDs will largely supersede photography for Earth-based imaging of all the planets. But to record Jupiter and all its moons still requires the larger field of a camera system, and indeed this can be done with a long telephoto lens (400 mm or more) mounted piggyback on a driven telescope. With a fairly fast film, try exposures of between ten and fifty minutes – the image of Jupiter must be overexposed in order to record the moons.

Radio Observation

Jupiter is a radio emitter, with strong interactions between Io and Jupiter's magnetic field, but those amateurs with radio telescopes will already know more about the techniques involved than there is space to repeat here.

Comet Impacts

Most amateur astronomers will have seen pictures of the spectacular effects produced when twenty or so pieces of the broken-up comet Shoemaker–Levy 9 (SL9) collided with Jupiter in July 1994. Indeed, some of the effects may still be observable when this book first appears in print. The resulting marks on the visible surface were so prominent that many could be seen through telescopes as small as 3-inch refractors.

If any marks are still visible, the usual observations should be performed: transits (preceding and following ends, as well as centres), drawings, darkness and colour estimates with and without filters, and latitude measurements.

Preliminary results already indicate that material from the impacts was thrown up to a great height, and that the comet had broken up into at least three types of nuclei – dense, solid objects of two different types, and lighter, more tenuous objects. Further analysis reveals a lot more about the composition of both Jupiter and SL9, which may or may not have been a typical comet.

Astronomers are now searching through records to see if there is any evidence of previous collisions. The *Voyager* photos of Callisto show a crater chain which could well be due to a series of impacts of an SL9-type object.

While the SL9 impacts were predicted well in advance, and thus widely observed, there may be other impacts in future of which we will have no warning.

Observers should, therefore, always be on the lookout for any unusual phenomena on Jupiter. SL9 produced obvious dark markings, but future impacts might have different results, depending on composition and where and how deeply they penetrate before exploding – the marks might be coloured, rather than dark, for example. One should also watch for any unusual flashes or fire-balls, either on the disc or on the limb – or, indeed, on the satellites.

In particular, it would be worth being always aware of the positions of any satellites, especially the inner ones, which are in eclipse but not in occultation, and watching for any sudden brightening into visibility resulting from fireballs from impacts on the far side of the planet. Satellite predictions are given in the *Handbook* of the BAA. Sudden brightenings of uneclipsed satellites should also be noted, but it is probable that very few impacts would produce fireballs brilliant enough to increase normal satellite brightness noticeably.

Any such phenomena which have been clearly and reliably observed should be reported immediately to a central coordinating body, such as the Jupiter Section of the BAA, or to *The Astronomer* or your local equivalent, in case they are the first of another series of impacts such as those of SL9.

If you have reliable records of any unusual observations in the past, you should report these too to the appropriate organisation. It is unlikely that there will be another comet impact in our lifetime, but it is well to be prepared, just in case. With Jupiter, always expect the unexpected!

Organisations and References

In the UK, the leading study group is the Jupiter Section of the British Astronomical Association (BAA). North American observers may join the relevant section of the Association of Lunar and Planetary Observers (ALPO).

The standard work reference work is *The Planet Jupiter*, by B. M. Peek, revised edition, published by Faber and Faber (London 1981). The latest detailed book is *Jupiter*, by R Beebe, Smithsonian Press (1994).

The BAA *Handbook*, published annually, gives all relevant details for Jupiter and its satellites. Similar information is available from ALPO.

Chapter 8
Saturn

A. W. Heath

Saturn is the outermost of the five bright planets known and observed since ancient times. However, it was not until around 1610 that astronomers realised that there was something rather different about this planet. It is possessed of a magnificent ring system which can be seen with a small telescope and a magnification of ×50 or so. We can see this ring system at all angles, from edge-on to about 29°, and open to both North and South directions. At edge-on, the rings disappear in all but the largest aperture telescopes.

The globe is not a perfect sphere; the polar diameter is about 11% less than the equatorial diameter. The equatorial 'bulge' is therefore greater than that of Jupiter. In many ways Saturn is not unlike Jupiter, in that it has dark belts and bright zones and the same nomenclature is adopted as for Jupiter. Spots are seen on the belts and in the zones from time to time, but they are usually short-lived. It is for this reason that our knowledge of the rotation periods for various latitudes is less accurate than it is for Jupiter.

Rotation is found to be 10h 14m at the equator, and transit observations have shown that spots in the higher latitudes reveal longer rotation periods. There is no sudden change of rotation period, as with System I and System II of Jupiter; instead it merely lengthens towards the poles.

Table 8.1 summarises the most important physical data about Saturn.

Saturn is less active than Jupiter and, while the spacecraft *Voyager* did find a small 'red spot', this feature can in no way be compared with the Great Red Spot

Table 8.1. Saturn: summary of data

Diameter (km); equatorial	120,000
polar	108,000
Rotation period; System I	10h 14m 00s
System II	10h 38m 25s
Inclination	26.73°
Magnitude at mean visual opposition	+0.7
Sidereal period (years)	29.458
Inclination to ecliptic	2.49°
Number of known satellites (1993)	18
Mean distance from Sun (AU)	9.5
(miles)	886×10^6
(km)	1427×10^6
Diameter as seen from Earth (arcsecs)	17 (approx.)
Radius of rings (km)	
Ring C, inner	74,000
Ring B, inner	91,900
Ring B, outer	117,400
Cassini's division, centre	119,000
Ring A, inner	121,900
Encke's division	133,400
Ring A, outer	136,600

of Jupiter. *Voyager* further revealed an extensive haze of some 70 km thickness which affects the visibility of the belts and zones. Wind velocities in the equatorial zone of the planet are the highest yet found in the atmospheres of any planet, including Jupiter.

Saturn is less dense than Jupiter; its density is less than that of water, so that if we could have an ocean big enough, it would float. It emits more energy than it receives from the Sun. The effective temperature is nearly 20°K higher than would be expected for the black-body temperature of an object at Saturn's distance from the Sun.

A useful aid to memory is the constant 'nine and a half'. Saturn's mean diameter is 9.5 times that of the Earth, it is 9.5 times as far from the Sun, its mass is 95 times that of the Earth and its year is 29.5 Earth years.

A convenience for observers of Saturn is that the planet rotates seven times in just under three days, so that features should be in nearly the same position at almost the same time every third day.

To the naked-eye observer, Saturn looks like a bright star with a softer and creamier lustre than either Venus or Jupiter. The brightness will depend on the extent to which the rings are open, as its mean visual magnitude is +0.7 with the rings closed but can reach –0.2 with the rings open. It is an easy target for photographers who may wish to record its position against the background stars, as it can be recorded in 20 seconds with an ordinary camera on a tripod and without driving mechanism.

A 3-inch refractor with a magnification of around ×50 will show the planet and its ring system, but an aperture of not less than 6 inches is needed for observations to be of value; ideally one should aim for an aperture of at least this size – the larger the better. It has been claimed that the best magnification for planetary detail is about equal to the diameter of the object glass or mirror in millimetres. To see the fine details of Saturn's belts and ring structure, a magnification of ×150 to ×300 is necessary, and therefore, according to the above rule, telescopes of 150 mm (6 inches) or more are clearly required.

The mean synodic period of Saturn (intervals between oppositions) is 378 days, therefore oppositions occur about a fortnight later each year. Oppositions vary considerably, however, in suitability for observations from a particular location, as the planet's declination varies between 26° north and 26° south. During the sidereal revolution of the planet, which takes 29.5 years, it is therefore usual for an observer at a particular location to experience relatively favourable apparitions for about fourteen years, followed by an equal period of unfavourable apparitions. Winter oppositions are the most favourable for observers in the northern hemisphere and are then least favourable for southern observers.

A further problem to observers of Saturn is the ring system itself. Except at the times when the rings are edge-on to us, they will obscure at least some part of the globe (see, for example, Figure 8.1). There are times when only one hemisphere is observable, so that any activity in the belts and zones is hidden and passes unrecorded. This is unfortunate, as our knowledge is therefore unavoidably incomplete.

For a given magnification, Saturn presents a much smaller disc than does Jupiter, being 19.5" for the equatorial diameter and 17.5" for the polar diameter. Observers should not be discouraged by this, as it applies to the globe only. The rings measure over 40 arc seconds from east to west, and contain much detail for the patient observer.

The determination of the longitude of surface features is a very important area of study. The transit

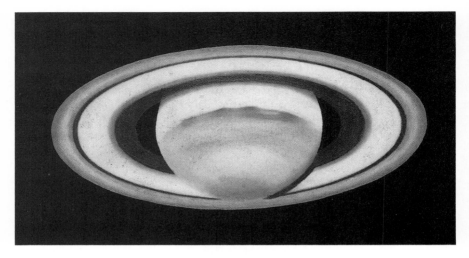

Figure 8.1
Saturn's rings, obscuring part of the planet's globe.

25 August 1958 1145 UT; 12-inch reflector ×300

method is used where a feature is timed when it crosses the central meridian. As with Jupiter, there are two systems. System I covers the two equatorial belts and the equatorial zone between, and has a period of 10h 14m (rotation rate: 844.3° per day). System II includes the rest of the planet with a period of 10h 38m 25s (rotation rate: 812° per day).

The nomenclature of the belts and zones is the same as that for Jupiter, but, due to the tilt of the planet, this can sometimes be confusing. Care must be exercised, as not all the belts or zones may be visible at any one time. A belt seen north of the North Equatorial Belt may not be the North Temperate Belt. If far north, it is likely to be the North North Temperate belt, with the North Temperate Belt absent. Determination of the latitude of each belt and zone is perhaps the best way of being certain of identity and the latitudes in Table 8.2 are given as a guide.

Long-term studies of the latitudes of belts and zones have shown that changes do occur from time to time.[1]

One of the main studies by the Saturn Section of the British Astronomical Association has been the making of intensity estimates of various features of the globe and the rings. Long term analysis of the results was carried out by Richard McKim and Keith Blaxall for the period 1943 to 1981[2] which was based on the work of 126 observers or observing groups, and over 80,000 intensity estimates were used.

The intensity scale is 0 to 10, where 10 is the black sky background and 1 is fixed as a bright reference, this being the brightest part of Ring B. The study itself was one of the most important carried out by the BAA Saturn Section, and it was found that both long-term and short-term changes in colour and intensity occur in the Saturnian atmosphere. Short-term intensity changes are often, but not always, associated with spot activity. Long-term changes can frequently be correlated with the seasonably variable solar altitude at a particular global feature. Increased solar heating or solar radiation intensity can darken a feature either by means of a photochemical mechanism, or by clearance of haze, allowing deeper views into the absorbing atmosphere. Increased solar heating may brighten the equatorial zone, possibly by raising the cloud top layer through the overlying haze. The fact that seasonal changes occur at all is sufficient evidence to show that the meteorology of Saturn is not merely controlled by released internal heat. All ring features exhibit intensity variations, and many of these can be correlated with the Saturnicentric latitude of the Sun. The Cassini division, Ring A and Ring C, all brighten up as the rings close up, due to partial bunching in line of sight. The higher particle density of the outer part of Ring B does not lead to a brightening of this ring as the Saturnicentric latitude of the Sun decreases. Instead, the intensity increases owing to the mutual eclipses of ring particles which can occur when the Saturnicentric latitude of the Sun is small.

With Saturn it is interesting to note that, in every Saturnian year, the northern hemisphere is exposed to the Sun for longer than the southern one. The belts and zones of the northern hemisphere are uniformly a little darker than their southern counterparts. Hemispherical inequality in both intensity and latitude is evident for both Saturn and Jupiter. The southern hemisphere is

Table 8.2. Latitudes of Saturn's belts and zones

Region	Latitude range
SPR	−60°
STB	−33° to −44°
SEB	−9° to −27°
NEB	+12° to +30°
NTB	+30° to +40°
NNTB	+56° to +62°
NPR	+60°

Note that southern hemisphere features are given a negative (−) latitude and northern hemisphere features are given a positive (+) latitude.

more active in the case of Jupiter whereas the northern hemisphere of Saturn has been the location of the most violent outbreaks on the planet, though we must not forget that a considerable amount of one hemisphere is hidden from view for much of the time.

The Ring System

The ring system makes Saturn such a splendid object, even when the rings are at a narrow angle. Three major rings are visible to the Earth-bound observer. The outer is known as Ring A, which is itself often divided into two parts by a faint dusky line known as Encke's division. The 'middle' ring is known as Ring B; it is the brightest ring and is separated from Ring A by a prominent division known as Cassini's division, so named after its discoverer. The inner part of Ring B is duller than the outer part. Inside Ring B is a further ring which is very dark and known as Ring C, which can be seen crossing the globe as a dark line. This is often confused with the shadow of the rings on the globe. A shadow of the globe on the rings causes them to have an apparent 'break', thus proving that the rings, like the planet, shine by reflected light from the Sun.

The ring system encircles the planet's equator, and shows itself at various angles, from edge-on to about 29° at maximum. This is due to a combination of the axial tilt of Saturn, which is 26.5°, and the 2.5° inclination of Saturn's orbit to the orbit of the Earth. On average, the rings take about 7.25 years to go from edge-wise to fully

Figure 8.2
Saturn's rings at near-maximum.

11 March 1989 0620 UT; 415-mm Dall–Kirkham ×237, seeing III–IV, transparency very good

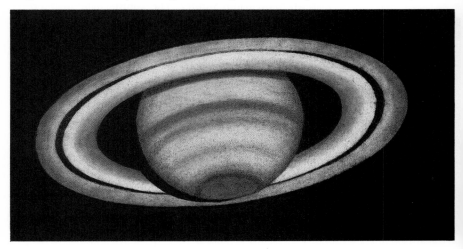

open, and this is the reason why Saturn varies in brightness to the naked eye observer. Figure 8.2 shows the rings almost at their maximum.

At intervals of 13.75 and 15.75 years, alternately, the Earth passes through the plane of the rings, and it is the eccentricity of Saturn's orbit which causes the difference. On each occasion the Earth can pass through the ring plane up to three times.

At times when the rings are edge-on, they are often invisible in small telescopes. This is because they are very thin, being not more than a few kilometres deep at most. It is at this time, or when the rings are at an angle of no more than a degree or so, that a truly magnificent sight awaits the observer, as any satellites which may be passing in front of them appear like tiny water droplets on a spider's web (see Figure 8.3). The invisibility of the rings may be due to them being exactly edge-on to the Earth, or exactly edge-on to the Sun with neither face being illuminated, or their unilluminated face being presented towards the Earth.

It is important to record the appearance of the rings at this time. Often, Cassini's division (the gap between Ring A and Ring B) is seen not as a dark line in the rings, as is more usual, but as two spots of light each side of the globe due to sunlight passing through it.

A hint of a faint ring outside the bright rings has been reported at times when the rings are near edge-on, as the ring material is thus concentrated in line of sight. At other times you would look straight through such a tenuous feature. *Voyager* confirmed the existence of further very faint rings outside Ring A, and these may well have been the cause of the earlier reports.

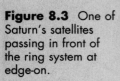

Figure 8.3 One of Saturn's satellites passing in front of the ring system at edge-on.

26 October 1966 2030 UT; 8-inch reflector ×274, seeing very good

Encke's division which, like Cassini's, is named after its discoverer, was considered by some to be a mere contrast effect between the two brightness levels of Ring A, but its true nature was revealed by *Voyager*. Other faint divisions can sometimes be seen, especially in Ring B, and any observer who detects such a division should note its position as accurately as possible by eye measurement and drawing. The position, or positions if more than one is seen, could be expressed as a fraction of the ring's width, e.g. 'two-fifths from the outer edge'. A comparatively large aperture and good seeing conditions are necessary to be sure of the reality of ring divisions other than Cassini's. Subjective impressions of divisions can be caused by poor seeing and small telescopes, and one must be aware that the differing levels of brightness in parts of a ring can cause the impression of a faint division at the point were the two levels meet. *Voyager* recorded hundreds of divisions in the rings, and also detail within the previously known Cassini and Encke divisions.

A comparatively little-known phenomenon of the ring system is a curious effect known as the *Bi-coloured Aspect* of the rings. It takes the form of a difference in the brightness of the eastern and western arms of the rings when viewed through coloured filters. Walter Haas, in New Mexico, USA, drew attention to this as long ago as 1949 when he noticed that the western arm of the rings was distinctly brighter than the eastern arm when seen with a Wratten No. 47 blue filter. No difference was detected with a red filter or indeed with no filter at all. Haas and others have seen this effect many times, it being visible with a variety of telescopes and eyepieces. However, in many views the arms do not appear to be different at all, each being equally bright both with and without filters. Occasionally the appearance has changed within only one or two hours. The effect has been photographed but, as yet, no one has succeeded in correlating the Bi-coloured Aspect with such parameters as the axial tilt of Saturn towards the Earth, the Earth–Saturn–Sun phase angle, or the position angle of the axis of Saturn in the Earth's sky. The effect would appear to be worthy of a more systematic investigation than it has yet received. Even if it turned out to be an illusion, we might well profit from learning the cause of that illusion. It is critical, of course, that visual observers make their observations totally independently of each other.

Faint, dark lines are sometimes seen in the rings. These features, known as 'spokes', have been seen from Earth and were considered to be optical effects rather than real features. *Voyager* found them to be real, so they have now become 'respectable'.

Occultations

Occasionally Saturn, its rings, or indeed both, may pass in front of a star. Occultations of bright stars can be very revealing, and variations in brightness give information, among other things, on particle density and the presence of faint divisions. The well known observation by Captain Ainslie on 1917 February 9 of the occultation of the star B.D. + 21°1714 by Ring A was final proof of the tranlucency of Cassini's division. The observation was a great credit to Ainslie, as the event had not been predicted.

A particularly interesting occultation occurred on 1989 July 3, in which a star was not only occulted by Saturn but also by the satellite Titan.[3] The star was the +5.8 magnitude gK4 type known as 28 Sgr. Modern technology has made new techniques available to amateurs, and this is well demonstrated by the observation of this occultation by C. Chrones and B. Duff in California. They used an 8-inch Schmidt–Cassegrain reflector, with a Javelin CCD video camera attached to it. A very detailed graph of the light curve was obtained. Later, two British observers, A. Hollis and R. Miles, paid special attention to the occultation by Titan. The event was well observed over much of western Europe, including the 'central flash' caused by Titan's atmosphere as it bent the star's light around Titan's limb and focused it. The whole event took some 5.5 minutes from first dimming to the full light of the star being regained. It has been calculated that such an event cannot occur more frequently than about once in 800 years.

Saturn not only occults stars, but can itself be occulted by our own Moon. Such events are interesting to watch, and it is important to time the events as they take place. They are not as revealing, however, as occultations of stars by the planet and rings, which can provide information about the ring structure and thus rank higher in scientific importance.

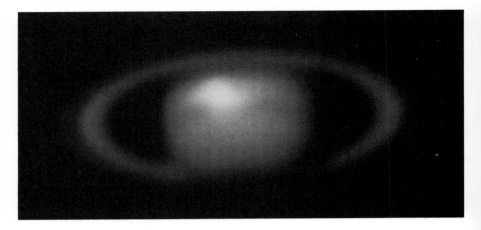

White ovals

Figure 8.4 The Great White Spot, a white oval which appeared in 1990.

2 October 1990
0939 UT

Large, white ovals have been seen on Saturn, the main ones being in 1876, 1903, 1933, 1960[4] and most recently in 1990. All occurred in the planet's northern hemisphere and most were discovered by amateurs.

Frank Melillo, writing in the Journal of the Association of Lunar and Planetary Observers (Vol 33, No. 7–9, July 1989[5]), drew attention to a possible periodicity of these ovals, as the dates are roughly 27–30 years apart. Melillo suggested that another white oval might be expected, and he was indeed right, for in September 1990 a most spectacular one appeared in the Equatorial zone. It was discovered by Stuart Wilber, an amateur astronomer of New Mexico, on September 25.[6] Figure 8.4 shows a photograph of the 1990 Great White Spot.

Caution is necessary, however, in the suggested 27–30 year periodicity. Table 8.3 gives the dates of discovery of both major and minor spots during the last 120 years. The spots of 1876, 1933 and 1990 occurred in

Table 8.3. Occurrence on Saturn of large, white ovals

Discovery date	Rotation period (centres)
1876 December 7 (Hall)	10h 14m 23.8s ± 2.5s
1903 June 15 (Barnard)	(in North Tropical Zone)
1933 August 3 (Hay)	10h 14m 15s
1960 March 31 (Botham)	(in North Temperate Zone)
1990 September 25 (Wilber)	10h 13m 48s

the Equatorial Zone, but those of 1903 and 1960 were at a higher northern latitude and less spectacular. The major outbreaks show a period of some 57 years, the lesser ones occurring in between. The major spots have similar rotation periods and all have occurred in the ecliptical longitude range of 290°–315°.

The appearance of the 1990 Great White Spot strengthens the hypothesis that such features appear on a periodic basis related to the position of the planet in its orbit, some time after midsummer in the northern hemisphere. The seasonal factor further supports the opinion that the meteorology of Saturn is not merely governed by released internal heating, but is also dependent on the level of incoming radiation.

The 1990 Wilber Spot may be compared with the 1933 Hay spot,[7] as both exhibited an overall drift in decreasing longitude, both were similar in size and appearance and both were better defined on the following side. Hay's spot became distended and ill-defined on the preceding side, as did the 1990 Wilber spot. Reference to the description of Hall's Spot of 1876 by A. F. O'D. Alexander indicates that it, too, was ultimately extended on one side into a streak. Sanchez Lavega *et al.*[8] note that the 1990 spot was rather brighter than the 1933 spot in visible wavelengths. A second spot was found following the main spot some 130° distant and there was a second spot some 80° following Hay's Spot too. Both the 1990 and 1933 spots interacted with the North Equatorial Belt, the breadth of which was considerably increased at the following apparitions in each case. It is reasonable to suppose that Hay's Spot and the 1990 Great White Spot owe their origin to the same process. Only time will tell if another Equatorial Zone spectacle of this magnitude will occur in or around the year 2047.

The Satellites

Saturn's satellites are of considerable interest. There are eight major satellites, the largest of which is Titan, together with a number of much smaller ones. Titan is of special interest in that it has an appreciable atmosphere; it compares in size with Jupiter's Ganymede and is therefore of planet size.

Most of the satellites follow a similar plane to Saturn's rotation, but Iapetus is more inclined and

Phœbe's orbit is retrograde. The high tilt of the planet's orbital plane means that satellite phenomena cannot be observed all the time, as in the case of Jupiter's Galilean satellites. It occurs only during the four or five apparitions centred on each passage of the Earth through the ring plane, the number increasing with proximity to the date of passage. In each sidereal revolution of the planet there are thus two periods of about ten years when no satellite phenomena can be observed. At this time they pass above or below the planet as seen from Earth. Only at times when the rings are edge-on or at a very small angle are the satellites occulted, eclipsed or in transit. The exception to this is in the case of Iapetus, which has an inclination of 15°; therefore such phenomena occur at a different time.

Titan is by far the largest of Saturn's family of satellites and takes a fraction under 16 days to complete an orbit of the planet. It has an almost circular orbit which lies in the orbital plane of Saturn, and it moves in the same direction that the planet rotates. Its substantial atmosphere exceeds that of the Earth by a factor of ten when considering the mass for a given area. The atmosphere has molecular nitrogen as its principal constituent, but it also has hydrogen-rich molecules like methane and the higher-order hydrocarbons. It may seem strange that this satellite, which has a magnitude of 8.3, appears dark when seen in transit in front of Saturn. It is so dark as to have been mistaken for its own shadow by some observers, and it is only when the true shadow appears that it is realised that it was Titan itself which had been seen earlier. Figure 8.5 shows both Titan and the shadow it projects on the globe.

The spectral variability of Titan has been investigated by the Unione Astrofili Italiani and the Saturn Section of the BAA. From 284 observations between 1977 and 1979 it was found that it was not possible to draw a light curve in relation to its orbital position.

Transits of satellites and their shadows are of special interest. Titan is quite easy, but the smaller satellites are far more difficult. Such phenomena are easier to observe if they fall on a light zone rather than a darker belt but, even then, good seeing and a fair aperture are required. Shadows of the more distant satellites are largely penumbral: that of Dione is some 25%. Barker, observing in 1936, found the shadow transits of Dione that had been predicted for August 22 and October 24 very difficult to see. During the 1966 apparition, Patrick Moore saw a transit of Rhea on August 1, and Rhea's

Figure 8.5 Titan, passing across Saturn's ring system, and projecting its shadow upon the globe.

29 October 1966 2140 UT; 8¹/₂-inch reflector ×274, seeing fair

shadow on August 10. The writer saw Rhea and its shadow in transit on September 19 and the shadow of Dione on September 20, both in good seeing with a 300-mm reflector.

Apart from Titan, the other satellites are much fainter, those of interest to the amateur falling between magnitude 10 and 14. Others are much fainter still and require much larger apertures to be observed. Hyperion, which is magnitude 14, can be located by using Titan as a guide. The orbital periods of Titan and Hyperion are almost 3:4, so that the two are in proximity at certain times. The brighter Titan can then be used as a guide to locate Hyperion.

Iapetus is of interest in that it is some two magnitudes brighter at western elongation than at eastern elongation. This satellite, which has a diameter of 1460 km, takes just over 79 days to orbit Saturn. The explanation for this difference is that the satellite has two vastly different hemispheres, though the boundary between the two is irregular. This appearance was confirmed by *Voyager*.

No new satellites were found until 1966, when a faint satellite of magnitude 14 was discovered photographically by Dr Dollfus. It has a near-circular orbit and a revolution period of almost 18 hours. Investigation by W. H. Julian of the BAA[9] of the satellite observations by Patrick Moore showed six occasions when the satellite may have been seen. These observations were made of course without a new satellite being expected. It would seem that this satellite, named Janus, was brighter than magnitude 14 as Patrick Moore used the 10-inch refractor at Armagh.

Table 8.4. Satellite data of interest to the amateur observer[10]

Satellite	Mean sidereal period (days)	Mean visual magnitude*
Titan	15.945	8.4
Rhea	4.518	9.7
Tethys	1.888	10.3
Dione	2.737	10.4
Enceladus	1.370	11.8
Hyperion	21.278	14.2
Iapetus	79.331	10.2–11.9

*The remaining satellites are all fainter than magnitude 14.5

The mean sidereal periods and visual magnitudes of seven of Saturn's satellites are given in Table 8.4.

While Saturn may not spring the frequent surprises that Jupiter does, each new apparition is watched with anticipation. Will Hay and Stuart Wilber had their patience duly rewarded, and no doubt others will too in the future.

Some years ago a friend of the writer, who was not an astronomer, asked if he could look at Saturn through a telescope. For several minutes he did not utter a word. He could not, for he stood with open mouth, consumed by the splendour of the spectacle before him. Indeed, such a thrill awaits all those wishing to share that experience.

References

1. Hollis A, 'Latitudes of Saturn's Belts – 1946–1976', *J. Brit. Astron. Assoc.* 91:1, p41 (1980)
2. McKim R and Blaxall K, 'Saturn – A Visual Photometric Study: Part 1', *J. Brit. Astron. Assoc.* 94:4, p145 (1984)
   'Part 2', *ibid.*, 94:5, p5 (1984)
   'Part 3', *ibid.*, 94:6, p249 (1984)
3. Hollis A and Mitton J, 'Titan/28 Sgr Occultation 1989 July 3', *J. Brit. Astron. Assoc.* 99:5, p205 (1989)
4. Alexander A F O'D, *The Planet Saturn*, Faber and Faber (1962)
5. Melillo F, 'A White Oval Watch on Saturn', *J. Association of Lunar and Planetary Observers*, 33:7–9, p138 (1989)

6. Heath A, 'Saturn 1990 – The Great White Spot', *J. Brit. Astron. Assoc.* 102:4, p210 (1992)
7. Heath A, '50th anniversary of discovery of Hay's Spot', *J. Brit. Astron. Assoc.* 93:5, p205 (1983)
8. Lavega S *et al*, 'The Great White Spot and disturbances in Saturn's equatorial atmosphere during 1990', *Nature* 353:6343, p400 (1991)
9. Heath A, 'Saturn 1966 – Report of Saturn Section', *J. Brit. Astron. Assoc.* 78:4, p284 (1968)
10. Taylor G, 'Satellites', *Handbook of the British Astronomical Association* (1994), p84

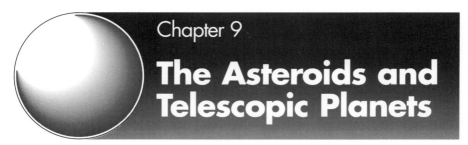

Chapter 9

The Asteroids and Telescopic Planets

Andrew J. Hollis

The Solar System contains many thousands of planetary bodies which are too faint to be seen by the unaided eye. All have been discovered during the last two hundred or so years. This group comprises the third and fourth largest of the Sun's planets (Uranus and Neptune – both about 50,000 km in diameter), the smallest planet (Pluto – approximately 2000 km in diameter) and several thousand asteroids (also known as the minor planets) which range in size from a few metres to almost 1000 km in diameter. In theory both Uranus and Vesta can be located without a telescope, though such sightings are very rare.

Only Uranus and Neptune appear large enough to show discs (about 3.5 and 2.5 arc seconds respectively) when viewed from the Earth. Both are gas giants and occasionally elusive dark cloud markings can be seen on the discs. Observing techniques used for Jupiter and Saturn are applicable, though great patience and care is required in their use. Many of the markings seen are contrast effects, but occasionally real atmospheric features can be detected.

At low magnifications both Uranus and Neptune appear stellar, and are best distinguished from background stars by reference to a good star atlas. The asteroids always appear as single points of light. Indeed the name asteroid means 'star-like', and was coined by William Herschel in 1802 shortly after the earliest discoveries. Observation of these bodies has many similarities with stellar astronomy, with the added complication that they do not remain in the same apparent places.

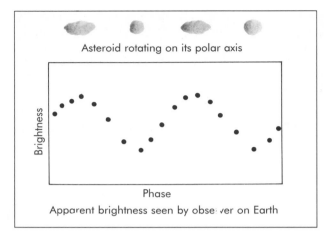

Asteroid rotating on its polar axis

Phase

Apparent brightness seen by observer on Earth

Figure 9.1 The rotation of an asteroid, plotted against its light curve.

Just locating an asteroid is an achievement, especially with the fainter examples. It is not possible to go to the eyepiece and say 'you have an asteroid' without some preparation. Armed with an ephemeris (table of predicted positions) and a star chart, the problem can be solved. Plotting their positions on a star chart will allow you to recognise the brighter asteroids immediately or else to narrow the search down to a few possible stars. Final recognition depends on determining which 'star' is moving. Either draw the field stars and wait for an hour or two, when the motion of the asteroid should be visible, or as a better alternative time the passage of the stars in the field across a wire.* Over the period of an hour the movement of the asteroid will reveal its identity as the transit time will have changed.

Having found the asteroid, its position can be checked. If this is done for a month or two and plotted on a star chart, it will serve as a practical demonstration of planetary motion.

Even relatively simple observations can be instructive. You can watch to see if the apparent brightness changes during the course of an evening as the asteroid spins on its axis. Figure 9.1 plots the rotation of an asteroid against its light curve. The really dedicated observer can contribute to fundamental research – one of the

---

* Use an eyepiece with a cross-wire (such as a guiding eyepiece). Centre the asteroid within the field, and switch off the drive. Time the interval when the stars drift behind the wire and repeat this an hour later. The interval will remain the same for the stars but will have changed for the asteroid (unless it is at its stationary point).

areas open to amateurs in observational astronomy. To be able to make such a contribution does mean using sophisticated equipment. CDD cameras, photographic telescopes or photoelectric photometers are all capable of precision work in the hands of a dedicated, careful observer.

Photometry

The apparent brightness of asteroids when viewed from the Earth is not constant. It changes by brightening as opposition approaches and fading as the Earth draws ahead after opposition. Observations made over a period of several months will show a change in brightness as the asteroid's distance from the Sun and Earth varies, and also as the phase angle (Sun–Planet–Earth angle; see Figure 9.2) alters. When the brightness estimates have been corrected to a standard distance a plot of brightness against phase angle can be made. The surface physical characteristics of the planet can be modelled using this plot.

The brightness will also vary during the course of an evening as the asteroid spins on its axis. These changes occur because the smaller members of the Solar System are usually irregular in shape. Many also have light and dark markings on their surfaces. The amplitude of the variations is usually quite small (typically 0.2 mag or less) though there are some examples which can exhibit a much greater range – for example, 216 Kleopatra varies by up to 1.6 magnitudes. If the polar axis points towards the Earth, no light variation will be seen as the

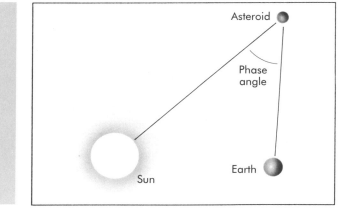

Figure 9.2 The phase angle – the angle between the Earth, the asteroid and the Sun.

rotation is in the plane of the sky. This may not be the case at a future opposition when the polar axis is directed away from the Earth.

An accurate value for the rotation period of an asteroid can be derived visually if the amplitude of the light curve is greater than about 0.4 or 0.5 magnitudes. Brightness estimates made at least every 20 or 30 minutes can be plotted on a light curve which can be used to determine the axial rotation period. If the amplitude is lower than this then more precise methods (using electronic equipment) are necessary to measure the brightness changes.

When planning a visual photometry project it is helpful to prepare a chart showing the path of the asteroid and identifying suitable comparison stars for estimating brightness. Such a chart for Uranus and Neptune appears annually in the BAA Handbook. Because the planets move relative to the background stars, a number of comparison stars strategically placed along the path should be chosen for this purpose. Whenever possible, comparison stars should be selected which have been measured photoelectrically so that their magnitudes are accurately known. When few such comparison stars can be found, additional stars can be taken from the AAVSO *Star Atlas* or the BAA's *Variable Star Fields*. Least reliable are stellar magnitudes taken from a star catalogue, as they tend to be relatively crude.

Several methods are available for estimating the visual magnitude of a minor planet; these are the same as those used by observers of variable stars. The three most commonly used approaches are the Fractional Method, Pogson's Step Method and the Argelander Step Method.

The Fractional Method is the simplest, and most beginners start in this way. Two comparison stars are used, one just brighter and one just fainter than the object. The brightness of the object under study is then estimated as a fraction of the difference in brightness of the two comparison stars.

Pogson's Step Method is relatively popular, even though it is probably the most difficult of the three methods. The observer trains himself (or herself) to recognise steps of 0.1 magnitude. The method demands great discipline from the observer, but can be used (with caution) when only one comparison star is present. Where possible, several comparison stars should be used.

The Argelander Step Method is perhaps the most satisfactory of the methods. The observer estimates the difference in brightness between the object and one comparison star. This process is then repeated several times using other comparison stars. If the star appears brighter than the planet for the same amount of the time that the planet appears brighter than the star, then the planet is assigned the same magnitude as that of the comparison star. If one appears brighter more often but occasionally appears fainter, then it is noted down as being one 'step' brighter. If one appears brighter most of the time, but is occasionally equal to the fainter, then it is two 'steps' brighter. The principle can be extended to several more steps with decreasing accuracy. Note that no attempt is made to estimate the size of an individual observer's 'step', which actually varies for different observers and at different times due to physiological factors. Typically a 'step' is in the range of 0.06 to 0.09 magnitude.

The visual magnitude of the object can be deduced from the estimates only when the comparison star magnitudes are known. Plotting the object's apparent magnitude against time will show if any regular or periodic fluctuations in brightness are taking place, or whether in fact the changes are solely random. If the regular change has an amplitude greater than 0.3 magnitude, then the double period (taking in two maxima and two minima) is most probably the true rotation period of the object. For the asteroids, typical periods lie between five hours and 20 hours.

Photoelectric and CCD photometry of asteroids may be carried out to investigate physical characteristics as well as rotation periods. This is the only way to study asteroids with low amplitude light curves. The writer has followed the rotation of 1 Ceres (amplitude 0.04 magnitude) using a 135-mm reflector and achieved a satisfactory light curve. The techniques used are specialised and beyond the scope of this chapter.

Astrometry

The simplest way to determine the position of an asteroid is to draw a sketch of it and the surrounding starfield. The position can then be determined by reading off the position from a star atlas or by calculating the position from the known positions of the stars in the field taken from a star catalogue. Much useful work was done in the early 1980s when the programme of recovery of lost asteroids was under way. Many quite bright (magnitude 11) examples had not been observed for some years and, if the orbits were accurately known, were well 'off-track'. At the time of writing (1993) only one remains lost (719 Albert, which was observed only during 1911). This method is not accurate enough for orbital analysis, but is a useful training to show how asteroids appear to move.

There are better methods available which make use of various types of micrometer. The filar micrometer is the most accurate, being capable of very high precision (0.2 arc second or even better). One of the simplest micrometers is the 'cross-wire' type and a description of its construction and use is given below. Generally, in making positional estimates of solar bodies, an accuracy of better than 30 arc seconds should be aimed for. This level of accuracy will indicate whether or not an asteroid is following its predicted path. If the body appears to be 'off-track', then photographic plates or CCD images should be taken in order to obtain a more precise positional determination (typically to better than 2 arc seconds).

Any pair of wires in the eyepiece can be utilised to determine the relative position of an asteroid. The wires do not need to cross at right angles, or even to cross at all. Any two straight, non-parallel wires or bars may be employed. Many observers will already possess an illuminated cross-wire guiding eyepiece which would be ideal for the purpose.

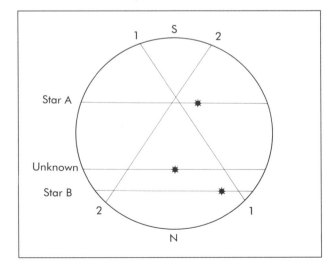

Figure 9.3 Using wires to determine the position of an asteroid, relative to known star positions. The wires do not need to be at right-angles, neither do they need to cross.

The aim is to determine the position of the asteroid (or other object) by measuring its position relative to stars of known position (see Figure 9.3). The *Hubble Space Telescope Guide Star Catalogue* (GSC) has been tested and proved to be accurate so, in practice, any star which happens to be visible in the telescope can be used. If the *Guide Star Caatalogue* is not available, then there are other astrometric catalogues which can be used, such as the SAO *Star Catalogue*. The reference stars can be plotted on a field drawing made beforehand and their positions obtained (the *Guide Star Catalogue* does not contain proper motions, so a correction cannot be made).

At the telescope, after locating the field, stop the drive and allow the stars to drift across the field of view. The times of transit of several reference stars and that of the object of study, across each of the wires, should be recorded as accurately as possible. Note that the absolute time (e.g. UT, GMT, TDT, etc.) is not required for this method. Instead, an accurate stop-watch (or event timer/logger) is all that is needed. Assuming that all timings are made in 'solar' seconds, the readings should be converted to 'sidereal' seconds by multiplying by the factor 1.002738. Note also that there is no theoretical objection to a considerable spread in the Right Ascension for the observed stars.

The raw timings have to be reduced to determine the celestial coordinates of the object under study. In the following example, the reference stars are designated A, B, C, etc. and the unknown is called X. The Right Ascension and Declination of each are referred to by $\alpha(A)$ and $\delta(A)$, etc. Lower case letters are used here to represent the time intervals in seconds (multiplied by the respective cosine of the Declination) between transits of the two wires for each star or for the unknown. The timings made at the telescope are denoted A1 and A2 for star A, etc.

Where stars cross the wires South of the intersection of the wires, the time interval is negative:

$$a = (A2-A1) \cos \delta(A), \ b = (B2-B1) \cos \delta(B), \text{ etc.}$$

and

$$x = (X2-X1) \cos \delta(X)$$

Note that since $\delta(X)$ is unknown, an approximation will suffice initially to allow $\cos \delta(X)$ to be evaluated.

The reduction of the observations is performed by means of the following equations, which are based on the comparison of similar plane triangles.

$$\delta(X) = \delta(A) + \frac{(a-x)(\delta(B) - \delta(A))}{(a-b)} \qquad \ldots(1)$$

and

$$\alpha(X) = \alpha(A) + (X2-A2) +$$
$$\frac{(a-x)}{(a-b)}(A2-B2) + (\alpha(B) - \alpha(A)) \qquad \ldots(2)$$

If $\delta(A)$ is almost the same as $\delta(B)$ the denominator will be small, so A and B should be chosen to be as far apart in Declination as is practicable. The values of $(a-x)$, etc., in Equation 2 should be divided by $15 \cos \delta$, where δ represents the Declination midway between those of the two objects concerned. However, as these factors appear in the numerator and denominator they will cancel out.

For minor planets situated low in the sky, the horizontal parallax and differential refraction should also be taken into account before the final reduction of the observations.

Using the cross-wires method, accuracies of about 0.5 seconds of time and 0.1 arc minutes have been obtained. The use of a third or even fourth reference star will increase confidence in the precision of the derived position. Further advantages of this method can be summarised as follows:

- An undriven altazimuth mount may be used.
- Illumination of the cross-wires is not essential.
- The balance of the telescope is not disturbed as the equipment is light.
- The equipment is simple, unsophisticated and relatively inexpensive.

Astrometry can be carried out photographically. This requires precise measurement of good quality images on negatives, or the computer reduction of on-line CCD data. Photography has now largely been superseded at larger observatories, and it seems likely that serious photographic astrometry by amateurs will be overtaken by CCD imaging. Photography is relatively expensive, and involves the use of chemicals and a darkroom for processing. A CCD image may need expensive equipment to obtain it, but the raw image can be used immediately and may be computer processed to improve its quality. Image measurement is carried out on-line and precise positions can be derived within minutes of obtaining the image – even more quickly if the hardware and software are located in the observatory.

Astrometry of near-Earth asteroids and the fainter bodies are specialist fields that do not attract enough serious attention. The serious amateur is able to obtain data that are valuable to assist in deriving accurate orbits for asteroids. The close approach of a small asteroid may only be visible for a few days, and all precision data are precious.

Occultations

From time to time members of the Solar System may appear to pass in front of a star or another Solar System body as seen from Earth. This only happens for the Sun during daylight, so solar occultations cannot be observed visually (though in theory an occultation could occur during a total eclipse, so it would be wrong to say they are never observable!). Solar occultations of radio sources, such as the Crab Nebula in Taurus, have been detected since the 1960s, but there are very few amateurs who possess suitable equipment to observe these. Lunar occultations occur frequently, and are considered in detail in Chapter 4 of this book.

This section describes observing the occultation of stars by asteroids. The principles are also applicable to occultations of stars by planets and their satellites, mutual occultations of one satellite by another (for example by Jupiter's Galilean satellites) and, most rarely seen of all, occultations of one planet by another.

The visibility track width is usually very narrow, as it is basically the same as the diameter of the occulting body. Since most asteroids are less than 200 kilometres in diameter there will be very few permanent observatories along the track. Many observers take portable telescopes and recording equipment to the predicted track to obtain their data. Every observation, whether positive or negative, is important in defining the limits of visibility, and hence the size and position, of the asteroid.

The dates of possible occultations are available from a number of sources. The most easily accessible of these is the January issue of *Sky & Telescope*, which gives predictions for that year. In recent years, the International Occultation Timing Association (IOTA) and the European Asteroid Occultation Network (EAON) have also been producing predictions.

Though the time of occultation can be determined accurately, the area of visibility can only crudely be estimated. Unfortunately, the orbits of most asteroids and the positions of fainter stars are not sufficiently precise to define occultation tracks accurately well ahead of time. The path could be in error by ±1000 km, or even more. The predicted times will be more accurate, and the probable error is unlikely to be more than a minute.

The prediction can be refined if astrometric plates are taken shortly before the expected event, to show both the asteroid and the target star. However, in most cases, this opportunity is not available. Predicted events may well be seen from a site even if the area of expected visibility does not include that site. Some that have not been predicted undoubtedly occur, and any observation would be fortuitous. Occultations provide the most precise information about the sizes and shapes of asteroids. As the precise velocity of the asteroid relative to the stars is known, the duration of each occultation gives an accurate chord length. If several observers obtain good timings, a picture of the projected shape of the asteroid at the time of the occultation can be derived. In addition, a position can be obtained with much greater precision than would be possible using more conventional methods of astrometry.

Concentrated monitoring of the 'fused' image of the asteroid and the star should commence about 10 minutes before the predicted time, and continue for at least the same period after it. At the instant of occultation the brightness will reduce by the predicted amount, as the light from the star is interrupted. The change in brightness depends on the relative brightness of the asteroid and the star being occulted. If the occulted star is a close double, and the brightness may appear to reduce (and reappear) in two steps – the finding of previously unknown double stars is a reality.

The important part of the observation is to obtain a time of the start and end of any occultation. An accuracy of 0.2 seconds, or even better if possible, is required. It is not sufficient to rely on the accuracy of a watch without having first checked it.

Several sources of accurate time are available for this purpose. The telephone speaking clock service (in the UK) is available twenty-four hours a day, and seems to be accurate to 0.05 seconds. Broadcast radio time signals are very accurate and to be recommended. WWV (in the USA), MSF (UK) or VNG (Australia) are the standard stations. At certain hours, domestic radio programmes give time signals which can be used for setting clocks. Teletext transmissions contain clocks, and these have been checked against radio time signals and appear to be accurate. Other sources of time, such as time-checks on the radio or television, are not sufficiently reliable and should be avoided.

If the time can be recorded automatically (e.g. by computer, chart recorder or cassette tape recorder), then absolute timing accuracy of better than 0.05 sec is possible. In this case, the operator's speed of reaction in seeing the event – the 'personal equation' – tends to the limit the final accuracy possible.

Care must be taken when observing not to miss the possible secondary events which may happen some time either side of the primary occultation. It was the observation of such events which led to the discovery of the ring around Uranus. Although termed 'secondary', such an event could produce a dip in brightness virtually identical to that of the primary event. Any secondary event, if recorded, may represent an occultation by an unknown satellite of the minor planet.

The photographer should drive his telescope as fast as, or slower than, the sidereal rate, to produce a trailed image. Field stars will appear as images of a constant thickness on the negative, provided the sky transparency remains constant. If an event occurs, the image of the occulted star will appear fainter during the occultation. Timings of the start and end of any occultation can be deduced from measurement of the lengths of the relevant parts of the trail. This assumes that the times of the start and finish of the exposure have been noted to the required degree of accuracy and that the drive rate remains constant.

The photoelectric observer will guide accurately on the field star, and ideally will superimpose radio time signals on the recording for accurate timing purposes. Reduction of the observation to derive accurate timings can be carried out later.

Discovery

In the 200 years since Ceres was found (Figure 9.4), almost 6000 asteroids have been discovered and their orbits established sufficiently accurately to allow them to be recovered at another opposition (either later, or earlier if photographic plates exist). There are as many known which have not had their orbits defined, and therefore have not been given numbers. Every year several hundred asteroids are allocated numbers and on

average 20 of these have been discovered by amateurs, mainly in Japan. Most of the 'new' asteroids are recoveries, where identification with earlier observations can be made.

UK amateur Brian Manning, who made several discoveries in 1989, was the first observer from Britain to discover an asteroid for 80 years. His discoveries were made photographically and were of 16th and 17th magnitude. It is unlikely that any asteroid brighter than this remains to be discovered, apart from a few small asteroids which may become bright for a few days as they make a close approach to the Earth.

The time has long passed when these minor bodies were considered merely as nuisances. Uranus and Neptune have been visited by the spacecraft *Voyager 2*, which has provided a snapshot of conditions on these remote worlds and their retinues of rings and satellites. The *Galileo* probe has returned pictures of 951 Gaspra and 243 Ida, on its way to Jupiter. Return visits or further exploration may take place in the future, but the precise format will be subject to confirmation when funding is approved.

For many years to come the main bulk of observation

Figure 9.4 Since Ceres (arrowed) was found more than 200 years ago, the orbits of almost 6000 asteroids have been established with sufficient accuracy to enable them to be found again.

7 March 1986 2030–2040 UT; RA 11h 20m, Dec + 20°, 45-mm f/3.5

will be carried by those resident on Earth. Observations described in this chapter will serve as an introduction to the serious study of the asteroids and remote planets.

As an aside, even counting asteroids (as others collect train numbers) has its devotees – the current record of visual sightings stands at about 1700 asteroids.

References

Cunningham C, *Introduction to Asteroids*, Willman-Bell (1988)

Hunt G and Moore P, *Atlas of Neptune*, Cambridge University Press (1994)

Hunt G and Moore P, *Atlas of Uranus*, Cambridge University Press (1987)

Kowal C, *Asteroids*, Ellis Horwood (1988)

Miner E, *Uranus*, Ellis Horwood (1990)

Moore P, *The Discovery of Neptune*, Praxis Press (1995)

Tombaugh C and Moore P, *Out of the Darkness: the Planet Pluto*, Stackpole Press and Lutterworth Press (1980)

Chapter 10

Meteors

Neil Bone

The observation of meteors remains a popular activity among amateur astronomers. This field has the obvious appeal of being accessible to those possessing only minimal equipment, and continues to throw up occasional surprises.

Meteor observing has, as yet, been touched little by the 'CCD revolution' which has brought changes to imaging technology in other areas. In part, the restricted sky area coverable using commercially-available chips makes these unsuitable for meteor work. Developments in low-light video camera technology and improved fine-grain photographic emulsions have, however, been embraced by meteor workers, and more participants are turning to short-wave radio as a complementary means of following shower activity.

The success of professional radio and radar work from the late 1940s onwards in monitoring activity led many amateurs to believe that they could no longer contribute usefully to the study of meteors. However, in common with a number of professionally-sponsored photographic programmes, such work has in many areas been discontinued.

It would be patently wrong to say that the patterns of meteor activity are completely understood and mapped out. The enhanced activity of the Perseids attending the return of their parent comet P/Swift-Tuttle in 1992, outbursts from the Lyrids and Ursids, and gradually increasing peak rates from the Geminids, are all witness to the dynamic nature of meteor streams. Changes may occur over a number of time-scales, necessitating continued coverage by observational techniques which have served well for many decades. Here, a review is given of the methods used by the Meteor Section of the British Astronomical Asscoiation (BAA), which share many common features with those employed elsewhere.

Visual Observation

By far the greatest bulk of amateur meteor work carried out comprises naked eye visual watches, which allow meteor rates to be determined, while also providing details pertinent to the physical characteristics of the meteoroids within streams. The practice of visual meteor plotting, a mainstay of amateur work up to the 1950s, has now largely been abandoned in view of the greater positional accuracy attainable by photography.

The equipment requirements for visual meteor work are minimal. In this work, the eyes' natural wide field of view is exploited, and no optical aid (beyond spectacles, if necessary) is used. The standard method is for the observer to be comfortably reclined on a deck chair or similar support, such that the sky at an altitude of 50° from the horizon may be watched. Observer comfort is an important consideration; an uncomfortable observer will not be concentrating fully, and will miss many of the meteors occurring, leading to a less reliable assessment of activity. It is important, in order that he or she stays alert, that the observer should be suitably protected against the elements. As in other fields, the use of a red torch to record by is essential for preservation of night vision. Spare batteries should, of course, be carried, along with a spare bulb. A hard-bound notepad or a clipboard with plenty of paper, and several pens and/or pencils, should obviously also be available. An atlas, such as the ubiquitous *Norton's Star Atlas*, is a further useful accessory.

As discussed later, visual data are normally reduced to obtain hourly rates of meteor activity. It is natural, therefore, that the minimum desirable length of a watch should be an hour. Weather conditions may not always allow this, of course, but watch intervals substantially less than an hour in length can prove unreliable as indicators of meteor shower activity. At the other extreme, observers should resist the temptation to carry out 'marathon' watches of several hours' unbroken duration. Breaks of 10–15 minutes should be taken to allow the observer a rest, refreshment, and to maintain alertness, after perhaps 2–3 hours' watch time during long stints.

Standard report forms are used for the recording of watch results by those bodies which collect meteor data. Details to be entered on the BAA forms are similar to those expected by other organisations world-wide. Watches should be logged and reported on separate sheets for each night – it is confusing in the extreme for the analyst to have to deal with several nights' data written up on a single form.

Another area where ambiguity can arise is in entering the date. To avoid this, meteor workers should, as standard, use a double-date system, as in 1994 November 15–16, regardless of whether the watch began before or after midnight. The times of the start and end of each watch interval, clearly listing any breaks, should be given, always in Universal Time (UT).

In the reduction of meteor results, one of the critical considerations is the state of the sky at the time of observation. Results are reduced with reference to a standard sky of limiting magnitude +6.5. In practice, such clarity and darkness can seldom be attained at most sites. Haze, thin cloud and light pollution may each play a part in degrading the sky, and thus reducing visibility. Obviously, if faint stars are not visible, neither will be faint meteors! Visual observers should take particular care to provide a record of the limiting magnitude during watches. The chart in Figure 10.1 presents the North Polar Sequence, used as a standard by BAA observers in estimating limiting magnitude. The limiting magnitude can change with time during a watch, and should therefore be checked regularly – perhaps every 30 minutes or so.

Under normal circumstances, there is little point in attempting visual meteor work under conditions where the naked eye limiting magnitude is worse than +5.0; the correcting factors applicable beyond this level become large, such that any derived rates are likely to have a high degree of uncertainty. Where unusual activity – such as potentially extreme high Perseid or Leonid rates – is possible, it may prove worthwhile to observe in conditions of bright moonlight or twilight, provided the limitations on accuracy in subsequent reductions are accepted.

Local obstructions or cloud may limit the area of sky visible. Again, note of these should be provided in the eventual report. Estimates are made to the nearest 10%. During the watch, the observer should record a number

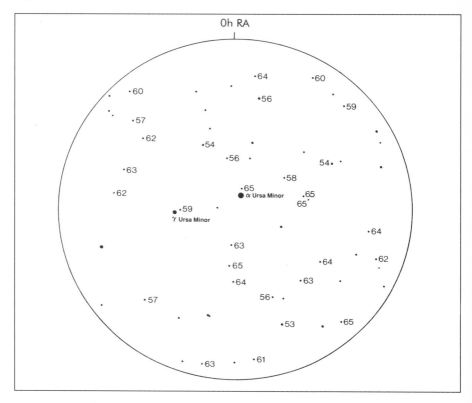

of parameters for each meteor seen. These are:

Time of appearance: Always in UT, normally to the nearest minute.

At times when photographic work is being conducted in parallel, an accuracy of ±3 seconds is desirable for bright events.

Magnitude: Meteor brightness should be estimated to the nearest whole magnitude (higher accuracy being difficult to attain for such short-lived events), using background stars or planets as comparisons. Table 10.1 provides a list of some suitable objects.

Type: Before commencing the watch, the observer should check the standard listings to see which showers are expected to be active on the night in question. Table 10.2 lists the principal showers.

More detailed lists can be found in the BAA *Handbook* and similar sources. From the tables, it is possible to determine the expected radiant position. As

Figure 10.1
The North Polar Sequence, used as a standard by BAA observers in estimating limiting magnitude.

a consequence of Earth's orbital motion, shower radiants appear to drift eastwards by about 1° per day, and allowance for this must be made if observing on nights away from maximum (the radiant positions given in shower listings are always for the time of maximum). For instance, at the beginning of August the Perseid radiant lies in Andromeda almost 1 hour in Right Ascension west of its location at maximum!

As an observational convenience, a radiant diameter of 8° is adopted for visual work. Meteors whose paths can be extended backwards across the sky to intersect the 8° circle centred on the expected radiant are taken to be shower members. To make meteor identification easier, the observer should select a field of view some 40°–60° in azimuth from the radiant.

Particularly in late July, when several showers may be active from radiants separated by only small angular distances, great care may be required when deciding upon shower association.

Meteors which do not appear to emanate from any known radiant expected to be active on a given night are recorded as *sporadics*. The sporadic background present throughout the year is, by its random nature, a useful control population against which to assess shower characteristics.

Constellation in which seen: BAA observers are asked to indicate in which part of the sky the meteor appeared.

Table 10.1. Some suitable comparison objects for assessing meteor magnitudes

Magnitude	Comparison object
−4	Venus
−2	Jupiter
−1	Sirius
0	Capella, Vega, Arcturus, Procyon, Rigel
+1	Altair, Deneb, Pollux, Regulus, Spica
+2	Alpha And, Beta And, Gamma And, Alpha Ari, Beta Aur, Beta CMa, Beta Cet, Alpha CrB, Gamma Cyg, Gamma Gem, Alpha Hya, Gamma Leo, Alpha Per, Alpha UMa, Eta UMa, Alpha UMi (Polaris), Beta UMi
+3	Theta Aur, Gamma Boo, Beta CMi, Beta Cyg (Albireo), Beta Dra, Epsilon Gem, Epsilon Leo, Epsilon Per, Zeta Tau, Delta UMa, Epsilon Vir
+4	Mu And, Beta Aql, Gamma Ari, Delta Aur, Kappa Cyg, Alpha Del, Gamma Del, Epsilon Del, Omicron Leo, Iota Peg, Eta Per, Chi UMa, Eta Vir
+5	Pi Cep, Gamma Com, 23 Ori, Tau Peg

Table 10.2. Principal meteor showers.

Shower	Activity Limits	Activity Maximum	Solar longitude	Max. ZHR*	Radiant position at maximum	Daily motion RA	Daily motion Dec	Notes
Quadrantids	Jan 1–6	Jan 3	283.1°	100	RA 15h 28m; Dec 50°			Strong only within 6 hours of peak. Bright members are often yellow-green or bluish.
Lyrids	Apr 19–25	Apr 21	032.1°	15	RA 18h 08m; Dec 32°	1.1°	0.0°	Fast meteors. Occasional strong returns, as in 1982.
Eta Aquarids	Apr 24–May 20	May 5	045°	35	RA 22h 20m; Dec –01°	0.9°	0.4°	Very fast meteors, often showing persistent trains. Best seen from southerly latitudes.
Alpha Capricornids	Jul 15–Aug 20	Aug 2	130°	5	RA 20h 36m; Dec –10°	0.9°	0.3°	Weak radiant, sometimes producing slow, bright meteors.
Delta Aquarids	Jul 15–Aug 20	July 29 / Aug 6	126° / 134°	25 / 10	RA 22h 36m; Dec –17° / RA 23h 04m; Dec 02°	0.8° / 1.0°	0.3° / 0.2°	Double radiant with separate maxima. Best seen from southerly latitudes.
Iota Aquarids	Jul–Aug	Aug 6	134°	10	RA 22h 10m; Dec –15° / RA 22h 04m; Dec –06°	1.1° / 1.0°	0.2° / 0.1°	Double radiant producing many faint meteors.
Perseids	Jul 25–Aug 20	Aug 12	140°	80	RA 03h 04m; Dec 58°	1.4°	0.1°	Excellent shower producing many fast, bright events with trains. Additional peak around solar longitude 139.5° has produced very high rates close to perihelion of Comet P/Swift–Tuttle in 1992.
Orionids	Oct 15–Nov 2	Oct 21	209°	30	RA 06h 24m; Dec 15°	1.2°	0.1°	Fast meteors, many with trains. Broad peak over several days, with sub-maxima.

Table 10.2. continued

Shower	Activity		Solar longitude	Max. ZHR*	Radiant position at maximum	Daily motion		Notes
	Limits	Maximum				RA	Dec	
Taurids	Oct 15–Nov 25	Nov 3	221°	10	RA 03h 44m; Dec 14°	0.8°	0.2°	Double radiant yielding slow, occasionally bright meteors over extended activity span.
					RA 03h 44m; Dec 22°	0.8°	0.2°	
Leonids	Nov 15–20	Nov 17	235.4°	See note	RA 10h 08m; Dec 22°	0.7°	−0.4°	Very fast, often trained meteors. In most years, ZHR around 15. Much higher activity around perihelion of Comet P/Tempel–Tuttle. Storms possible in 1998/1999.
Geminids	Dec 7–15	Dec 13	262°	100	RA 07h 26m; Dec 32°	1.1°	−0.1°	Excellent shower of slow, often bright meteors. Ideal for photography.
Ursids	Dec 19–24	Dec 23	271°	10	RA 14h 28m; Dec 78°	0.9°	−0.5°	Poorly-observed shower. may produce outbursts.

* ZHR = Zenithal hourly rate.

This can be useful, for example, in identifying bright events seen – or even photographed – from several locations.

Train: Particularly among the brighter events (produced, typically, by the largest and/or highest velocity meteoroids), there may be a tendency for the production of a short-lived streak of light along the meteor's path, visible for a matter of seconds following extinction of the meteor itself. These *persistent train* phenomena are of interest, giving some information as to the nature of incoming meteoroids, and of the state of the atmosphere in the meteor layer at 80–100 km altitude. Records of the presence, and duration in seconds, of such trains should be made. Trains with very short durations (less than 0.5 second) are described as *wakes*. Rarely, very long duration trains may be seen to distort and move under the influence of high-atmosphere winds, and their decay may be followed using binoculars. Sketches of such trains' development are of interest.

Other notes: In addition to the preceding essential details, observers may wish to record other information, such as pronounced colour (normally only found with the brighter events), flares or fragmentation in flight, and apparent speed.

Many observers restrict their meteor work to those times when the richest showers – the Perseids, Geminids or Quadrantids – are expected to be active. Much valuable work remains to be done on other showers, however, and more dedicated meteor enthusiasts will use every opportunity to cover these.

Visual meteor work can be most satisfying, particularly if the observer is treated to good activity, allowing reasonable numbers of meteors to be logged. In order that such results should have more than just personal value, they should of course be reported to the BAA Meteor Section or equivalent national body for inclusion in subsequent analyses.

Analysis

Since the 1960s, the emphasis in visual meteor analyses has shifted to the determination of activity profiles which can, in turn, yield information on the distribution of meteoroids within streams. Magnitude data allow this distribution to be considered with respect to meteoroids of differing sizes. For instance, around the maximum of the Geminids, there is a progressive increase in activity of bright meteors, produced by larger particles; sorting by the Poynting–Robertson effect has led to larger meteoroids being more common in the outer parts of the core filament in the Geminid stream, such that these are encountered later in the Earth's passage through it.

An indication of the particle size distribution is given by the *population index* (*r*), which relates the numbers of meteors in adjacent magnitude intervals. *r* may be determined from watch data, using meteors observed in the magnitude range from about 0 to +3 (the probability of missing meteors in the +4 to +5 range causes the linear increase in numbers at fainter magnitudes to break down). Typical values for major showers are of the order of 2.25. A small population index typifies a greater proportion of bright meteors, a large value indicating high proportions of faint events – large and small meteoroids respectively. For the sporadic background throughout the year, a value of *r* = 3.42 may be used.

The population index is a fundamental factor in determining corrected Zenithal Hourly Rates (ZHR). Clearly, a stream producing large numbers of faint events (such as the Leonids) will be adversely affected by poor limiting magnitude (LM) conditions. Corrections are applied as $r^{(6.5-LM)}$.

The standard Zenithal Hourly Rate used to express shower meteor activity is a theoretical index of the activity which would be expected for a single experienced observer, under perfectly cloudless skies with LM +6.5, and with the radiant overhead. In practice, LM will be poorer than +6.5, and the radiant altitude will vary during the night. When the radiant is low, fewer meteors will be seen as a result of several effects, including loss of meteors among haze and obstructions near the horizon. As a reasonable approximation, these factors may be allowed for by 1/sin *a*, where *a* is the radiant altitude. *a* can be calculated for the mid-watch interval from:

$$\sin a = \sin \delta \sin \phi + \cos \delta \cos \phi \cos H$$

where δ is the radiant's Declination, φ the observer's latitude, and H the radiant's Hour Angle. H is determined from Local Sidereal Time minus Right Ascension of the radiant.

Table 10.3. 1991 Perseids: individual observations

Date: 1991 Aug 11–12 Observer: N. Bone, Chichester, UK. Lat. 50°49.8' N., 0°48.3' W.

Start	End	Dur^n (hr)	Mid-watch	Av LM*	Av Cl*	F	Spor	CHR	Per	Alt rad	Perseid ZHR
2150	2250	1.00	2220	5.0	10%	1.11	4	27.8 ± 13.9	11	32.6°	73.6 ± 22.2
2300	0000	1.00	2330	5.25	10%	1.11	3	15.5 ± 8.9	11	39.7°	50.1 ± 15.1
0000	0100	1.00	0030	5.25	–	1.00	3	14.0 ± 8.1	14	46.5°	56.2 ± 15.0
0100	0200	1.00	0130	5.5	–	1.00	6	20.5 ± 8.4	9	53.8°	26.2 ± 8.7
0200	0300	1.00	0230	5.5	–	1.00	11	37.6 ± 11.3	20	61.7°	53.4 ± 11.9

Example: Perseid ZHR for 2300–0000 is derived as follows:

$$1/\sin 39.7 \times 2.35^{(6.5-5.25)} \times 1/(1-0.1) \times 11$$

$$1.566 \times 2.91 \times 1.11 \times 11 = 50.1$$

Error $\sqrt{11}/11 \times 50.1$

Sporadic corrected hourly rates are calculated allowing only for LM and obscuration. For 2300–0000, therefore, we have:

$$3.42^{(6.5-5.25)} \times 1.11 \times 3 = 15.5 \pm 8.9.$$

Table 10.4. 1991 Perseids: BAA recorded data

Date: 1991 Aug 11–12 BAA Averages

Mid UT	Tot H	Av LM	Av F	Spor	CHR	Per	Alt rad	Perseid ZHR
21h 43m	4.83	5.55	1.01	17	11.7 ± 2.8	33	30.2°	31.0 ± 5.4
22h 35m	10.03	5.43	1.03	33	12.5 ± 2.2	88	34.8°	39.3 ± 4.2
23h 38m	7.20	5.41	1.03	27	14.8 ± 2.8	62	41.2°	34.3 ± 4.4
00h 31m	7.17	5.54	1.03	32	14.9 ± 2.6	102	47.9°	44.7 ± 4.4
01h 34m	8.75	5.39	1.01	62	27.7 ± 3.4	119	55.0°	43.0 ± 3.9
02h 28m	4.77	5.60	1.02	24	15.7 ± 3.2	71	61.4°	37.1 ± 4.4

Radiant altitude figures are based on the average location of the observers.

Sky obscuration (by cloud, buildings, and so on) is allowed for by $1/(1-c)$, where c is the decimal fraction of sky hidden. c can be taken as an average for each hourly interval.

Finally, correction should be made for duration of the watch interval, if this differs from 1 hour, by $1/H$, where H is the actual observed time in hours.

Statistical error is given by $(\sqrt{N}/N) \times$ ZHR, where N is the number of meteors observed.

Putting these factors together, we may now apply these to some actual observations from the 1991 Perseids, in Table 10.3. Population index $r = 2.35$ is used for the Perseids.

Clearly, those values based on larger meteor samples have a smaller likely error. The accuracy of the ZHR calculation can be improved by averaging together results from several observers. BAA results from the same night are presented in Table 10.4. The value of pooling together results from several experienced observers should be self-evident.

The mean ZHR figures in Table 10.4 show much less scatter than those derived from a single observer, and are fairly typical of the Perseids' activity level some 24 hours ahead of their regular maximum. Interestingly, an abrupt increase occurred about 6 hours later, indicative of material recently released into the Perseid meteor stream by Comet P/Swift–Tuttle – a feature seen again from 1992–94.

ZHR figures may be plotted against time, as in Figures 10.2 and 10.3. Figure 10.2 shows the Perseids' profile in 1983, a 'normal' year, where activity rose steadily to a sharp peak around Aug 12–13. In Figure 10.3, however, the presence of a more concentrated filament of débris in the stream produces an additional, higher peak almost a day earlier.

Activity from year to year may be more directly compared in terms of a time system based on solar longitude, which gives a direct indication of Earth's orbital position. In general, where a meteor stream is fairly stable, the peak ZHR will occur around the same solar longitude from one year to the next. Since Earth takes 365.25 days to arrive back at the same position, however, the regular Perseid maximum at solar longitude 140.0° occurs roughly 6 hours later in UT each year. Where radical changes, such as the appearance of the novel peak in the Perseids, have arisen, the difference in activity around the same solar longitude can become striking. Comparison of Figures 10.2 and 10.3 clearly

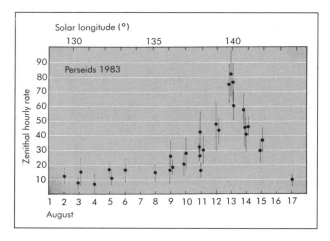

Figure 10.2
ZHR figures for the Perseids in a 'normal' year, plotted against time.

demonstrates the appearance of the novel peak around solar longitude 139.5° – absent in 1983 – while the normal peak at solar longitude 140.0° is similar in both 1983 and 1993.

Photography

Meteor photography offers the chance to obtain permanent records of the brighter events, and has more or less completely supplanted visual plotting as a means of obtaining accurate positional data. While casual photographers, seeking only to record meteor trails for æsthetic purposes, may prefer to use a basic 35-mm camera equipped with a wide-angle lens, serious workers will be more rigorous in their equipment selection. To be of use for measurement purposes, photographic exposures need to be timed accurately; start and end times of exposure (and, preferably, times of appearance of any recorded meteors) should ideally be quoted to within 3 seconds. A reasonably long focal length is invaluable in providing a larger image scale, improving the accuracy of subsequent measurements.

In its Photographic Programme, the BAA Meteor Section recommends the use of 35-mm cameras equipped with 50-mm lenses operating at $f/2$ or faster, or medium-format (120) cameras such as the inexpensive Lubitels, which have a fixed 75-mm $f/4.5$ lens. Cameras may be static on a tripod (or, if a battery of several is in use to increase sky coverage, on a common mount), or driven at sidereal rate. Best capture rates are enjoyed with cameras aimed around 50° elevation, some

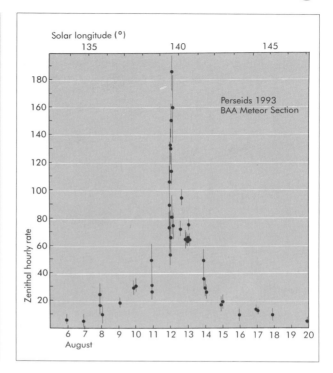

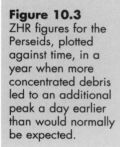

Figure 10.3
ZHR figures for the Perseids, plotted against time, in a year when more concentrated debris led to an additional peak a day earlier than would normally be expected.

20°–30° in azimuth from an active radiant. Several years' work suggests Ilford HP5-plus film, uprated to 800 ISO, to be the most suitable emulsion for meteors. Exposure times are governed largely by sky brightness. Where light pollution is significant, 10 minutes may be the maximum possible. At dark sites, exposures up to perhaps 25 or 30 minutes are possible. It should be borne in mind that extended exposures run the risk of crowding out meteor trails among the long star trails in undriven frames (see, for example, Figure 10.4).

An ever-increasing problem for meteor photographers is the presence of artificial satellites, which can leave meteor-like trails across the frame. Reliable meteor identification on the eventual negative is facilitated by the use of a rotating shutter in front of the camera(s). A shutter generating 15 breaks per second will 'chop' the fast-moving trail of a meteor, leaving satellites as solid trails, thereby allowing the two to be distinguished. Figure 10.5 shows an example of a meteor photographed using such equipment.

Meteor photographs are useful in two principal areas. Firstly, since only those events brighter than magnitude +1 are recorded, the capture rate provides an

index of bright meteor activity as a sub-population in a shower; this may vary from the visual profile. Secondly, measured positions allow assessment of the size, position and motion over time of shower radiants, when several independently-recorded trails are combined.

Multi-station recording of individual events may be used to triangulate meteors. Where at least one of the stations operates a rotating shutter, it is possible to determine the velocity of the incoming meteoroid and, further, derive its original solar system orbit.

More specialised areas in meteor photography include spectrography using objective prisms, and fireball patrol work using 'fisheye' all-sky lenses. Both of these, perhaps as a consequence of the relatively low capture rates, tend to be rather neglected by amateur observers, but valuable work is certainly possible in either field..

A handful of workers now use low-light video camera detectors, rather than film, to obtain real-time coverage. While cost is certainly a consideration, the increased yield (meteors to magnitude +5 can be recorded with a 28-mm $f/2.8$ lens), coupled with the ability to review the results at leisure, may be considered advantageous. It seems likely that such work will become increasingly popular in the future.

Figure 10.4
Long-exposure photographs using undriven equipment may result in meteor trails being crowded by the long trails of stars.

Figure 10.5 A rotating shutter will break up the fast-moving trails of meteors, thus avoiding the problem shown in the previous figure.

Telescopic Observation

Standard naked eye watches of the type described above allow detection of meteors to a limit of magnitude +5 or so. The use of binoculars allows meteors to perhaps magnitude +9 to be recorded. While faint meteors should, in principle, be more common, the restricted field of view (typically 5° for 10 × 50 binoculars) limits the observed rates, such that telescopic work is the preserve of the dedicated observer and is unlikely to attract casual participants. Valuable results can, however, be obtained.

Telescopic meteor work remains an area where plotting can be employed to good effect. Pre-drawn charts for selected sky areas (available to participants in the BAA Meteor Section's Telescopic Programme) are used, and the accuracy of meteor plots by experienced observers is sufficient to allow shower radiant positions to be determined, Several new minor radiants have been discovered by telescopic observers.

Wide-field binoculars or a 'rich field' telescope are the preferred equipment for this work. The instrument should be steadily mounted such that the observer can watch in relative comfort. Watches are typically conducted in stints of 15–20 minutes, separated by regular breaks; the lower eye-relief resulting from viewing limited fields can be tiring for the observer. Two separate, nearby fields are usually watched alternately.

The start and end times of each watch interval should, naturally, be recorded. As in naked eye work, a record of the field limiting magnitude is made. Meteor plots are numbered in succession as they are made, and other details may be recorded as appropriate (principally time of appearance and magnitude). Telescopic meteors may be described as type aa (entirely within the field), type ao (beginning in the field, ending outside), type oa (beginning outside, ending within the field), or type oo (crossing the field from outside). Shower association cannot normally be ascribed at the time of observation due to the limited field of view.

Radio

A number of amateur observers employ fairly standard receivers to detect meteors by means of forward scatter. In such work, the aim is to detect signals from a VHF radio transmitter lying far beyond the receiver's normal horizon, by virtue of their reflection from the brief trail of ionisation produced by meteors. Ideally, the receiver should be fed from a 4-element Yagi antenna. Counts of the 'pings' resulting from forward scatter by meteors occurring between transmitter and receiver give some indication of the activity level. Most radio meteors detected correspond to events in the very faint, sub-visual range.

A distance of 1000–1500 km between transmitter and receiver is regarded as optimal, with meteors being detected around the mid-point of the transmitter-receiver path. So, for example, UK operators seek to detect meteors occurring over Germany by virtue of forward scatter from Eastern European transmitters. Suitable transmitters, broadcasting on the 4-metre band from 66–70 MHz, may be identified from the *World Radio TV Handbook*.

In order that forward scatter counts can be reduced to give an indication of absolute rates, corrections need to be applied to account for two factors. Where shower activity is under study, the sporadic background should be determined for a number of days on either side of the expected shower dates, and subtracted from the total. The raw counts will also be influenced by the diurnally-changing geometry of the transmitter-receiver path relative to the shower radiant, introducing an observability function: most activity will be detected when the radiant lies around 90° in azimuth from the direct line joining receiver and transmitter.

Provided allowance is made for these factors, forward scatter radio allows the determination of shower activity under cloudy or day-lit conditions, and may be seen as an important addition to the range of coverage attainable by the serious observer.

Bibliography

Texts at amateur level devoted to meteor work are comparatively rare. Up-to-date reports of recent activity are often given in the *Journal* of the British Astronomical Association, and in magazines aimed at the more serious amateur, such as *Sky & Telescope*; a few examples are listed below.

Bone N, *Observer's Handbook: Meteors*, George Philip (1993)

Brown P and Rendtel J, 'Shooting Stars: The 1994 Perseids', *Sky & Tel.*, 89:1, 108–110 (1994)

Duffett-Smith P, *Practical Astronomy with your Calculator*, Cambridge University Press (1983)

Elliott A J and Bone N M, 'Video observations of the Geminid meteor shower in 1990', *J. Br. Astron. Assoc.*, 103:4, 181–183 (1993)

Evans S J, 'Meteor photography', *J. Br. Astron. Assoc.*, 102:6, 336–342 (1992)

Evans S J and Bone N M, 'Photographic and visual observations of the Geminid meteor shower in 1991', *J. Br. Astron. Assoc.*, 103:6, 300–304 (1993)

Evans S J and Ridley H B, 'The spectrum of a Perseid meteor', *J. Br. Astron. Assoc.*, 103:1, 27–29 (1993)

Kronk G W, *Meteor Showers: A Descriptive Catalog*, Enslow (1988)

Norton A P, *Norton's Star Atlas*

Roggemans P, *Handbook for Visual Meteor Observations*, Sky Publishing (1989)

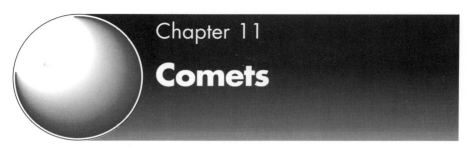

Chapter 11

Comets

Jonathan Shanklin

Comets are highly unpredictable, which is one aspect which makes them so interesting to observe; their appearance can change from night to night, and each comet is different. Around twenty comets are usually discovered or recovered each year; of these, perhaps four will come within the range of amateur instruments.

Comets can be classified into three groups: short period comets (those with periods less than 20 years), intermediate period comets (those with periods between 20 and 200 years) and long period comets (those with periods greater than 200 years). The long period comets include those in parabolic and hyperbolic orbits which will never return.

Several sub-groups of families of comets can be identified. The comets of the Jupiter family, which have aphelia near the orbit of Jupiter, were captured into short period orbits by the planet. The Kreutz sungrazing group may have originated in the break-up of a single large comet over 10,000 years ago. Comet Encke is associated with several asteroids and meteor showers and possibly comet Rudnicki may be the largest remaining fragment of another large comet which broke up several thousand years ago.

Sometimes an object which is initially identified as an asteroid develops a coma. Depending on the circumstances it may be reclassified as a comet (e.g. 1990, UL3 = P/Shoemaker–Levy 2) or retain its asteroidal designation (2060 Chiron). The new numbering scheme aims to reduce such confusion in the future.

For the Beginner

Sources of Information

In order to observe a comet you must know that it exists, and where it is in the sky. A list of cometary positions is known as an ephemeris. The two main sources of information that provide rapid notification of the discovery, position and brightness of comets are the circulars of the British Astronomical Association (BAA) and the TA electronic circulars. The BAA's Comet Section newsletter and BAA circulars will usually have location charts for comets, but you will need to plot positions in a star atlas unless the comet in question is a very bright object.

Star Atlas

You will need a good star atlas, showing stars at least as faint as the comet you plan to observe. The best one to get is the atlas of the American Association of Variable Star Observers (AAVSO), as this has comparison star magnitudes suitable for estimating a comet's brightness down to around the 10th magnitude. Others include *Norton's Star Atlas* (useful for plotting the overall track of a comet and selecting stars to hop from) or *Uranometria 2000.0 Vols 1 & 2*. One thing to watch out for is the epoch of the atlas and the comet positions: they must be the same. Positions for Epoch 2000 are fine for *Norton's 2000* and *Uranometria*, but the AAVSO atlas (which is the standard reference atlas) is Epoch 1950. The Section circulars will continue to use Epoch 1950 until an updated version of this atlas is introduced and the majority of observers are using it. Computer based planetarium or star charting programs can enable you to make your own finder charts, though you should be cautious about using the magnitudes from such programs unless no other source is available.

Star Hopping

If your telescope doesn't have setting circles, the way to find objects in the sky is by a process known as star hopping. First, find a convenient naked-eye star close to the object. Centre this in the finder or in the main telescope, using the lowest magnification you can. Compare the field with a detailed star atlas and decide on the orientation of the field. Now all you have to do is move from star to star until you come to the object you are seeking. A faint comet can sometimes be easier to spot if you move the telescope backwards and forwards a little, once you have found the field.

Instruments

Although a few comets every decade may be visible to the naked eye, some form of optical aid is required to see the majority. In general, a short focal length and low magnification are desirable for observing comets, though a few are seen better with a higher magnification.

No single instrument can be used to observe all comets, so a compromise is needed. Simple 7×50 binoculars might be the best instrument for a beginner, though they will only show the brighter comets down to around magnitude 8. If you later find that comet observing doesn't suit you, you still have an instrument that you can use in several other fields of astronomy. To see fainter (more typical) comets you will need a telescope or large binoculars and, if possible, a light-pollution-free site. Using a telescope at a typical UK site, the approximate magnitude limit (for a condensed comet) is shown in Table 11.1.

The faintest comet that you will be able to see will be two to three magnitudes brighter than the faintest star that you can see with the same telescope. You should always use the lowest magnification and smallest aperture that will clearly show the comet for the magnitude estimate; you may, however, need a larger aperture and higher power to see detail in the inner coma. A reflector of short focal length on a Dobsonian mounting is perhaps the best choice, unless you want to make photographic observations.

Table 11.1. Comet magnitude limits for given telescope apertures

	Aperture (mm)					
	50	80	150	200	300	400
Faintest magnitude (approx.)	8.0	9.5	10.5	11.0	12.0	13.0

Other Equipment

You should also have:

- A notebook to record your observations in, a pen or pencil to write and draw with, and a 'plastic' rubber. If you later add extra details (e.g. the calculated coma diameter or comet magnitude), use a different coloured pen to record the new information.
- A watch or clock set to UT (GMT) to record time to the nearest minute. A watch with a stop-watch function and dual time is ideal. If normal summer time (BST) is in operation you need to subtract one hour to get UT.
- A dim, red light to allow you to see the star atlas and write your notes, but not so bright that it destroys your dark adaptation. A cycle rear lamp is a little on the bright side, unless the batteries are on the way out. Spare batteries are often needed.
- Warm clothing. It can get quite cold at night, even in the summer months in a sheltered observatory. You should aim to wear several layers of clothing for best thermal insulation. You should not drink or smoke if you plan to make observations.

Visual Observation

Most comets are not very spectacular, just faint smudges of light similar in appearance to many deep sky objects. This similarity prompted Messier to produce his famous *Catalogue*, which was designed to help him to discover new comets. The average comet appears as a small patch of light, a few minutes of arc across (a distance similar to the separation of the Galilean satellites from Jupiter), possibly brightening a little towards the centre. This patch of light is known as the coma and is produced by the outgassing of material from the pile of icy rubble a few kilometres across that is the true comet nucleus. The nucleus is far too small to be seen through even the largest of telescopes, but you may see a false nucleus or disc-like nuclear condensation at the heart of the coma. The head of the comet consists of the coma and the nucleus. A bright comet may develop a tail, or even tails, which generally point away from the Sun. Figure 11.1 shows the parts of a comet.

Unfortunately, prospects for visual observation in the UK, particularly of the fainter comets, are not

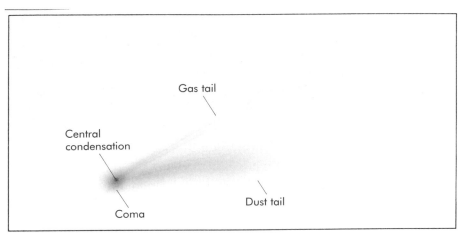

Labels in figure: Gas tail · Central condensation · Dust tail · Coma

Figure 11.1 The parts of a comet.

encouraging because of the widespread increase of light pollution. A dark site helps you to make good observations: if you can't see a magnitude 7 comet in 7×50 binoculars you should try to find a darker site. The higher the site the better, as this puts you above more of the atmosphere so that there is less scattering and absorption of light.

You should keep a record of all your observations in a book. It is best to have a rough notebook to use at the telescope, since inevitably it will get damp and dirty, and a finished version in which to complete your notes and drawings. Your book can record extra details not reported on the standard forms, e.g. your personal narrative. You should always record your observations as you go, rather than committing them to memory and writing them down later. It is worth recording regularly-used details in the front of the book, e.g. the latitude and longitude of your observing site, the magnification and field diameter of various eyepieces, the separation of the standard double stars, etc.

You will need to become properly dark adapted before starting to observe. During this time you can write down some background information in your observation book. Record the name and the official designation of the comet which you are about to observe, and where you are observing it from. Next, write down the year, month and day, using the double date (e.g. 1995 March 22/23) to avoid confusion. It is important to record the time of the observation; for visual cometary work, accuracy to the nearest 15 minutes (0.01 day) is sufficient. Table 11.2 shows the decimal day (the proportion of the day so far elapsed) for a given time (UT).

The appearance of a comet very much depends on the type, aperture and focal ratio of the instrument, on the type and magnification of the eyepiece being used, and on the observing conditions; so these facts need to be recorded. A comet generally appears fainter with higher magnification, larger aperture and poorer conditions. To find the comet, plot its position on a suitable star atlas and then star hop or use setting circles to find the field. For a faint, diffuse comet, accurate positions are essential, as the object may be barely observable above the sky background. Here it may help to print a computer generated chart showing the comet and faint stars prior to the observing session.

Estimating the Magnitude

It can be very difficult to make good magnitude estimates of comets, because they are often large and dif-

Table 11.2. Decimal day from UT hour and minute

Hr.	00	05	10	15	20	25	Min. 30	35	40	45	50	55	60
0	0.00	0.00	0.01	0.01	0.01	0.02	0.02	0.02	0.03	0.03	0.03	0.04	0.04
1	0.04	0.05	0.05	0.05	0.06	0.06	0.06	0.07	0.07	0.07	0.08	0.08	0.08
2	0.08	0.09	0.09	0.09	0.10	0.10	0.10	0.11	0.11	0.11	0.12	0.12	0.13
3	0.13	0.13	0.13	0.14	0.14	0.14	0.15	0.15	0.15	0.16	0.16	0.16	0.17
4	0.17	0.17	0.17	0.18	0.18	0.18	0.19	0.19	0.19	0.20	0.20	0.20	0.21
5	0.21	0.21	0.22	0.22	0.22	0.23	0.23	0.23	0.24	0.24	0.24	0.25	0.25
6	0.25	0.25	0.26	0.26	0.26	0.27	0.27	0.27	0.28	0.28	0.28	0.29	0.29
7	0.29	0.30	0.30	0.30	0.31	0.31	0.31	0.32	0.32	0.32	0.33	0.33	0.33
8	0.33	0.34	0.34	0.34	0.35	0.35	0.35	0.36	0.36	0.36	0.37	0.37	0.38
9	0.38	0.38	0.38	0.39	0.39	0.39	0.40	0.40	0.40	0.41	0.41	0.41	0.42
10	0.42	0.42	0.42	0.43	0.43	0.43	0.44	0.44	0.44	0.45	0.45	0.45	0.46
11	0.46	0.46	0.47	0.47	0.47	0.48	0.48	0.48	0.49	0.49	0.49	0.50	0.50
12	0.50	0.50	0.51	0.51	0.51	0.52	0.52	0.52	0.53	0.53	0.53	0.54	0.54
13	0.54	0.55	0.55	0.55	0.56	0.56	0.56	0.57	0.57	0.57	0.58	0.58	0.58
14	0.58	0.59	0.59	0.59	0.60	0.60	0.60	0.61	0.61	0.61	0.62	0.62	0.63
15	0.63	0.63	0.63	0.64	0.64	0.64	0.65	0.65	0.65	0.66	0.66	0.66	0.67
16	0.67	0.67	0.67	0.68	0.68	0.68	0.69	0.69	0.69	0.70	0.70	0.70	0.71
17	0.71	0.71	0.72	0.72	0.72	0.73	0.73	0.73	0.74	0.74	0.74	0.75	0.75
18	0.75	0.75	0.76	0.76	0.76	0.77	0.77	0.77	0.78	0.78	0.78	0.79	0.79
19	0.79	0.80	0.80	0.80	0.81	0.81	0.81	0.82	0.82	0.82	0.83	0.83	0.83
20	0.83	0.84	0.84	0.84	0.85	0.85	0.85	0.86	0.86	0.86	0.87	0.87	0.88
21	0.88	0.88	0.88	0.89	0.89	0.89	0.90	0.90	0.90	0.91	0.91	0.91	0.92
22	0.92	0.92	0.92	0.93	0.93	0.93	0.94	0.94	0.94	0.95	0.95	0.95	0.96
23	0.96	0.96	0.97	0.97	0.97	0.98	0.98	0.98	0.99	0.99	0.99	1.00	1.00

fuse; practice helps. The observed visual magnitude is dependent on a number of factors, including how much defocusing you use, and the telescope aperture and magnification. You should use the smallest aperture and lowest magnification that clearly shows the comet.

To make the estimate, the technique is similar to that used for a variable star estimate, where you choose (from the AAVSO atlas or a variable star chart) one comparison star (A) that is a little brighter than the variable (V), and another that is a little fainter (B). If the variable is closer in brightness to the brighter star then your estimate might be A 1 V 2 B (here the range is divided into three parts). If it is about midway, estimate A 1 V 1 B (the range is divided into two parts). If the variable is closer in brightness to the fainter star, esti-mate A 2 V 1 B (the range is divided into three parts again). If the variable is much closer to the brighter star your estimate might be A 1 V 3 B (the range is divided into four parts), but it is rarely necessary to divide the range into more than four parts. The brighter star is always given first. If the variable appears identical in brightness to A, then record it as =A. If you can't see the variable at all, just record the faintest comparison that you can see, for instance if you can see B but not the variable, then record <B or [B.

To estimate for a comet, the same procedure is fol-lowed, but you have to put the comparison stars out of focus. Select some likely comparison stars, either on the AAVSO atlas or from a nearby BAA or AAVSO variable star field. For fainter comets it may be preferable to use the North Polar Sequence. Ideally the comparison stars should be in the same field, but this is rarely possible; if they are not you will have to memorise the brightness of the comet and may have to alternate between the fields several times.

First, centre the comet in the field and try to memo-rise its appearance. Next, choose one comparison star that is a little brighter and another a little fainter than the comet, and defocus (racking the eyepiece in) until they appear the same size as the focused comet. Are they still brighter and fainter? If not, choose different comparisons. When you are happy, make the estimate in exactly the same way as for a variable star. This tech-nique is known as the Sidgwick (S) Method.

As an example, suppose you had used the variable star T UMa (on the BAA's VSS program), and made the estimate: P (1) comet (2) Q (i.e. the comet is closer in brightness to P). P is magnitude 9.85, Q is magnitude

10.29, the difference between them is 0.44; 1/(1+2) of this is 0.15, so the comet is 9.85 + 0.15 = magnitude 10.00 (however, we only record to 1 decimal place, so the result is magnitude 10.0).

If you have time, repeat this for another pair of stars, because there is often some discrepancy in the magnitudes of the comparison stars from different fields, and several independent estimates will improve the accuracy of the magnitude determination.

Always give the source of the comparison star magnitudes, but there is no need to record the actual estimate on the report form, unless you do not have a suitable catalogue or atlas. In this case show the stars you have used by means of a field sketch on a separate sheet.

Submit your observations to the appropriate organisation in monthly batches. The best time to do this is during the full Moon period, when many comets will be invisible.

Estimating the Coma Diameter

As the comet approaches the Sun, sublimation of ices normally increases and the coma expands. If the comet approaches very close to the Sun, the coma may appear to shrink again, though this is probably due to increasingly poor observing conditions. The apparent size of the coma depends on the distance from the Earth and the true diameter of the coma. Measuring the diameter of the coma thus gives some physical information about the comet but also gives a comparison between different observers' sky conditions.

The simplest way to measure the coma diameter is to compare the coma with a pair of stars of known separation. It is easiest to use two stars in the same field as the comet. Their separation can be measured later from an atlas or star chart, calculated from the stars' positions or found by using a planetarium computer program. For a quick estimate the stars Mizar and Alcor form a convenient yardstick; the separation of Mizar and Alcor is 11.8 arc minutes, or Alcor to its magnitude 8 companion is 6.3 arc minutes.

Advanced Observation

Telescope Magnification

When fully dark adapted, the pupils of most people's eyes dilate to give a pupil diameter of around five to seven millimetres; younger people generally have larger diameter pupils when dark adapted. The best magnification to use is one which gives the same exit pupil as your own pupil diameter. To find this magnification, simply divide the aperture of your telescope by the pupil diameter of your eye. As an example, if your telescope is a 200-mm reflector and your pupil diameter is 5 mm, the optimum magnification is ×40. Two eyes are better than one, so 7 × 50 or 11 × 80 binoculars are commonly used for comets brighter than 10th magnitude; these have exit pupils of seven millimetres. For fainter comets many observers use telescopes in the range 150–300 mm, with magnifications of ×20 to ×100. Some comets can be seen more clearly when a much higher magnification than the optimum value is used.

Magnitude Estimate

Never use a filter while making a magnitude estimate.

There are several alternative methods of making a magnitude estimate. The Morris (M) Method is a slight extension of the Sidgwick Method, where the comet is defocused just enough to give it a uniform brightness. Its brightness is then memorised and the comparison stars are defocused further until they are the same size as the defocused comet.

In the Bobrovnikoff (B) Method both the comet and the comparison stars are defocused together, until the stars are about the same size as the defocused comet. The estimate is then made by direct comparison between the comet and the selected stars.

Finally, in the Beyer (E) Method, the comet is defocused until it disappears; this technique is good for bright comets.

The Sidgwick Method is good for magnitude estimates of diffuse or faint comets and poor skies. The Bobrovnikoff and Morris Methods are better for well condensed comets (Degree of Condensation – DC > 6); using the Bobrovnikoff Method you may have to defocus to 2 or 3 times the coma diameter, so it is not so good for large comæ. When defocusing reflector telescopes you should rack inwards if possible, to avoid vignetting.

Try not to check the magnitude of the comet against the ephemeris before you start observing. This can subconsciously bias you, so that you observe what you expect to see. This sometimes means that you miss a cometary outburst or a sudden fading. Don't force your observations to agree with the ephemeris magnitude, which can often be out by several whole numbers. As the comet's total magnitude depends on the surface brightness and size, a large diffuse comet which is apparently faint can actually be quite bright. Some comets, discovered with large telescopes, are initially assigned visual magnitudes that are too faint; comet McNaught–Russell 1993 v was a case in point – expected to be magnitude 15, it was found to be magnitude 10. It is worth attempting observation of some of these fainter comets just in case they are brighter than predicted. It is also worth observing in poor conditions – for example, comet Tanaka–Machholz 1992 d had an outburst during the full moon period. It is also worth comparing your observations with those made by experienced observers. Yours should vary in the same way, though it is quite common to have a systematic difference.

Because the observed magnitude depends on aperture and magnification, it is best to try to use the same combination throughout an apparition, provided that the range in the brightness of the comet is not large. When you do need to change to a different telescope, make observations with both instruments for a couple of weeks so that you get a good overlap. This helps to determine the systematic effects caused by such changes.

To encourage proper methodology, the *International Comet Quarterly* (ICQ) is planning a project to observe selected Messier objects; these will probably include M78 and M53 for binoculars and M66, M84, M86, M81, NGC 3640 and NGC 4147 for telescopes. Comparison star charts for these fields will be issued by ICQ as part of the project.

Star Magnitude Sources

The use of good comparison sequences is important. The recognised sources and their codes are given in Table 11.3; these include the AAVSO atlas (AA), the

North Polar Sequence (NP), variable star charts which use photoelectric sequences (AAVSCO – AC, BAA – VB), the *Fourth Yale Bright Star Catalogue* (YF) and, for fainter comets, the *Hubble Guide Star Catalogue* (HS). Of these, NP is perhaps best as it can be used for most comets (unless you have a telescope with an English equatorial mount). It is always a good idea to check several stars in the sequence and to use several different comparison stars. Beware of using some older variable star charts which often use 'guesstimated' magnitudes.

Table 11.3. Magnitude reference sources and codes

Code	Source
Standard	
AA	AAVSO Variable Star Atlas.
AC	Variable star charts of the AAVSO.
HS	V Magnitudes from the Hubble Space Telescope astrometric catalogue of stars on compact disc.
NP	North Polar Sequence; three charts published by the AAVSO showing stars with useful range mag 5 to 17.
S	Smithsonian Astrophysical Observatory Star Catalogue.
SC	Sky Catalogue 2000.0 (stars of mag <8.1).
SP	Skalnate-Pleso Atlas Catalogue (Atlas Coeli catalogue).
TC	Tycho catalogue from Hipparcos (to be published in 1996).
VB	Variable star charts of the BAA.
Alternative	
AE	Planetary or stellar magnitudes from the Astronomical or Nautical Almanac (for use with bright comets).
CS	Catalogue of Stellar Identifications (1979, Strasbourg). Large compilation of many catalogues. The visual magnitudes with colons (:) should be avoided if possible.
LN	Lampkin's Naked-Eye Stars.
TB	Supernova Search Charts, by G. D. Thompson and J. T. Bryan Jr. (1989, Cambridge University Press).
Y	Yale Bright Star Catalogue (early editions).
YF	Yale Bright Star Catalogue (fourth edition should be used in preference to earlier editions).
Do not use	
UA	Atlases Borealis, Eclipticalis or Australis.
UM	Magnitudes of galaxies, nebulæ, etc.
UN	Norton's Atlas.
UP	Any standard photographic atlas (e.g., Falkauer, Stellarum).
US	Skalnate Pleso Atlas.

Note: additional codes are given in the full ICQ magnitude reference list

Table 11.4. Coma diameter from transit time and declination

Comet dec.	1	2	3	4	5	6	7	8	9	10	20	30	40	50	60	70	80	90	100
											Transit time (seconds)								
0	0.3	0.5	0.8	1.0	1.3	1.5	1.8	2.0	2.3	2.5	5.0	7.5	10.0	12.5	15.0	17.5	20.0	22.5	25.0
5	0.2	0.5	0.7	1.0	1.2	1.5	1.7	2.0	2.2	2.5	5.0	7.5	10.0	12.5	14.9	17.4	19.9	22.4	24.9
10	0.2	0.5	0.7	1.0	1.2	1.5	1.7	2.0	2.2	2.5	4.9	7.4	9.8	12.3	14.8	17.2	19.7	22.2	24.6
15	0.2	0.5	0.7	1.0	1.2	1.4	1.7	1.9	2.2	2.4	4.8	7.2	9.7	12.1	14.5	16.9	19.3	21.7	24.1
20	0.2	0.5	0.7	0.9	1.2	1.4	1.6	1.9	2.1	2.3	4.7	7.0	9.4	11.7	14.1	16.4	18.8	21.1	23.5
25	0.2	0.5	0.7	0.9	1.1	1.4	1.6	1.8	2.0	2.3	4.5	6.8	9.1	11.3	13.6	15.9	18.1	20.4	22.7
30	0.2	0.4	0.6	0.9	1.1	1.3	1.5	1.7	1.9	2.2	4.3	6.5	8.7	10.8	13.0	15.2	17.3	19.5	21.7
35	0.2	0.4	0.6	0.8	1.0	1.2	1.4	1.6	1.8	2.0	4.1	6.1	8.2	10.2	12.3	14.3	16.4	18.4	20.5
40	0.2	0.4	0.6	0.8	1.0	1.1	1.3	1.5	1.7	1.9	3.8	5.7	7.7	9.6	11.5	13.4	15.3	17.2	19.2
45	0.2	0.4	0.5	0.7	0.9	1.1	1.2	1.4	1.6	1.8	3.5	5.3	7.1	8.8	10.6	12.4	14.1	15.9	17.7
50	0.2	0.3	0.5	0.6	0.8	1.0	1.1	1.3	1.4	1.6	3.2	4.8	6.4	8.0	9.6	11.2	12.9	14.5	16.1
55	0.1	0.3	0.4	0.6	0.7	0.9	1.0	1.1	1.3	1.4	2.9	4.3	5.7	7.2	8.6	10.0	11.5	12.9	14.3
60	0.1	0.3	0.4	0.5	0.6	0.8	0.9	1.0	1.1	1.3	2.5	3.8	5.0	6.3	7.5	8.8	10.0	11.3	12.5
65	0.1	0.2	0.3	0.4	0.5	0.6	0.7	0.8	1.0	1.1	2.1	3.2	4.2	5.3	6.3	7.4	8.5	9.5	10.6
70	0.1	0.2	0.3	0.3	0.4	0.5	0.6	0.7	0.8	0.9	1.7	2.6	3.4	4.3	5.1	6.0	6.8	7.7	8.6
75	0.1	0.1	0.2	0.3	0.3	0.4	0.5	0.5	0.6	0.6	1.3	1.9	2.6	3.2	3.9	4.5	5.2	5.8	6.5
80	0.0	0.1	0.1	0.2	0.2	0.3	0.3	0.3	0.4	0.4	0.9	1.3	1.7	2.2	2.6	3.0	3.5	3.9	4.3
85	0.0	0.0	0.1	0.1	0.1	0.1	0.2	0.2	0.2	0.2	0.4	0.7	0.9	1.1	1.3	1.5	1.7	2.0	2.2

Coma

Measuring Coma Diameter

Measuring the coma diameter is quite important, since it not only tells us how the comet is behaving, but it also says something about the observing conditions at the site. There are three main ways of measuring coma diameter:

- The simplest way, as noted earlier, is to compare the coma with a convenient pair of stars in the field, though for brighter comets Mizar–Alcor form a convenient yardstick. For fainter comets, estimate the diameter relative to two field stars – their separation can be measured later from an atlas or star chart, calculated from the stars' positions or found from a planetarium computer program. For reference, the separation of Mizar–Alcor is 11.8 arc minutes, or Alcor to its 8th magnitude companion is 6.3 arc minutes.
- With the telescope drive stopped, time how long the comet takes to transit across a wire or the edge of the eyepiece field with a stop watch (alternatively, you could time how long a star takes to move the same distance). The coma diameter = 0.25 × (time of transit, in seconds of time) × cos(declination). Table 11.4 gives values for a range of times and comet declinations. The measurement should be repeated several times and an average taken.
- Estimate the diameter relative to the eyepiece field width. You still have to use the second method to measure the eyepiece field width, but this only needs to be done once.

Degree of Condensation (DC)

The way the light distribution within the coma varies is called the degree of condensation; it should be given a whole number value between 0 and 9 (do not use half steps, or give a range of values rounded up or down). Table 11.5 describes the values, which are illustrated in Figure 11.2.

A comet which is completely diffuse, with no apparent change in intensity across the coma, has a DC of 0. A comet which appears as a star-like point or disc has a DC of 9. DC 5 is when the intensity has dropped to half

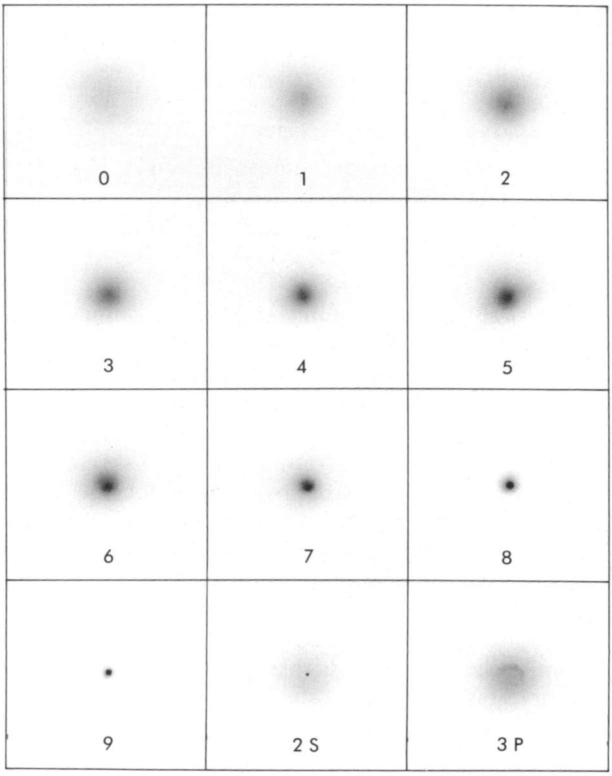

0	1	2
3	4	5
6	7	8
9	2 S	3 P

Figure 11.2
Degree of condensation (DC); these are the values described in Table 11.5.

at half the coma diameter. When an otherwise diffuse coma has a nuclear condensation, a false nucleus, which presents a step in the brightness distribution, estimate the DC by averaging these features. A comet which brightens a little towards the centre, but which also has a star-like false nucleus, might have a DC of 3. To describe the comet fully, add 'D' after the DC value if a disc-like nuclear condensation is present, 'S' if a star-like nucleus is present, 'N' if the condensation is nearly star-like, and 'P' if there is a discontinuity or step in the brightness distribution. In general, use only one letter; if the feature is faint, use a lower case letter instead.

A true nuclear condensation is rarely seen, and is possibly caused by an outburst; if so, its diameter should increase from night to night, and should be measured. The magnitude of this condensation (or the nuclear magnitude) can be recorded, but will generally only be of indicative use. Early observers thought that this star-like feature was the real nucleus, but the real nucleus is far too small to be seen in a telescope. Beware of the confusion between the terms 'nuclear condensation' and 'central condensation'; the latter merely indicates that the coma is much brighter towards the centre. A condensed comet need not have a nuclear condensation.

Features in the Coma

The physical appearance of the comet will depend on sky conditions; a bright sky drowns out much of the coma and the tail structure. Appearance also depends on the relative proportions of dust and gas in the coma. In general, a predominantly gaseous coma will appear less condensed than one predominantly composed of dust. Most of the light of the gaseous coma comes from C_2 Swan emission bands which lie between 430 and 620 nm, whereas the dust coma merely reflects the solar continuum. When a comet is more than 2.5 AU from the Sun, the C_2 emission tends to be very weak or absent, so only the dust coma is seen.

The Lumicon comet filter is based on the Swan bands and therefore enhances a gassy comet. The coma diameter and DC, etc., will change when viewed through the filter, and you should always report what you observe without the filter. If there are significant changes when the comet is viewed through a filter, these should be reported using the 'Comments' section on the report form. Some filters absorb so much light that faint comets still remain invisible. Because the spectral sensitivity of the eye changes as it dark adapts, it is possible that the appearance of the coma will also change.

Table 11.5. Degrees of condensation

Value	Description
0	Diffuse coma of uniform brightness
1	Diffuse coma, with slight brightening towards centre
2	Diffuse coma, with definite brightening towards centre
3	Centre of coma much brighter than edges
4	Diffuse condensation at centre of coma
5	Condensation appears as a diffuse spot at centre of coma
6	Condensation appears as a bright diffuse spot at centre of coma
7	Condensation appears as a fuzzy star that cannot be focused
8	Coma has virtually disappeared
9	Stellar or disc-like in appearance

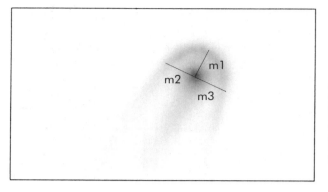

Figure 11.3
Measuring a comet's hood.

Detail can sometimes be seen in the inner coma, but physiological and psychological factors often play a part and we 'see' the defects in our own eyes. Very few comets show features in the coma. John Bortle has proposed the following classification scheme for features that may be seen.

H (Hoods or envelopes): These are only seen in bright comets near perihelion, and are more often photographed. They are the largest features to be seen. If you do see a hood, some simple measurements can give clues on the rotation of the nucleus (see Figure 11.3): measure the distance from the centre of the coma to the leading edge of the hood (m1) and to the edge of the hood in the two directions perpendicular to this (m2 and m3: they may not be the same in each direction).

R (Rays of the ion tail): These rarely extend all the way to the false nucleus, instead usually ending a few minutes of arc away. They are blue, can be seen with the Lumicon comet filter, and are quite common.

S (Shadow of the nucleus): This is a rare apparition, and needs a 40-cm or larger telescope to be seen easily.

F (Fountains): Around 1 arc minute high in the Sunward direction, these are suppressed by a Swan band filter. They are diffuse features, subtending an angle of anything from a few to 100°, and are more common than jets.

J (Jets): These are the rarest features. Very fine and delicate, they are always curving and possibly spiral, rapidly changing and less than 60 arc minutes high. They need a large aperture in order to be seen.

Because the nucleus of a comet is no more than a fragile pile of rubble, many comets have been observed to fragment. The most well known example, of course, is comet P/Shoemaker–Levy 9, which collided with Jupiter in July 1994. Sometimes this process is visible in amateur instruments – for example, in March 1976 the nucleus of comet West was observed to split into a number of fragments, and in September 1994 components A and D of comet P/Machholz 2 were easily visible in the same low-power field. What is often seen is the formation of several nuclei which gradually separate from the main nucleus, fade and eventually disappear. Many comets also undergo outbursts, and these can considerably distort the light curve.

There are a number of techniques that can be used for drawing the features of a comet. These include:

- Pseudo isophotes. Approximate contours of equal brightness are drawn to illustrate the coma and the tail. This is a good technique for beginners, as it requires little artistic ability
- Negative shading (the usual technique). Rub a pencil on a sheet, then use a finger or a piece of tissue paper to transfer the graphite to the sketch pad; a paper stump can be used to add tone, or an eraser to subtract it. You can use a finger to smear the sketch as necessary to achieve graduation of contrast
- Stipple. This is the best method for the non-artist. Lots of little dots are made using a fine felt tip, or self inking, pen to portray the brightness of the coma.

A scale bar (e.g. 1 inch in length) should be included in the drawing, and North should be marked. It is best to make a sketch and notes at the eyepiece, going on to complete the drawing indoors in comfort. Beware of bias, and of drawing what you expect to see.

Comet Tails

The majority of comets don't have significant tails. Those that approach close enough to the Sun, however, may develop a gas tail (Type I) or a dust tail (Type II). The gas tail is usually straight, blue and filamentary, and may suffer disconnection events, though these are rarely seen visually. The dust tail can be curved, and is yellow from reflected sunlight. What is seen by the observer is very dependent on viewing geometry; when

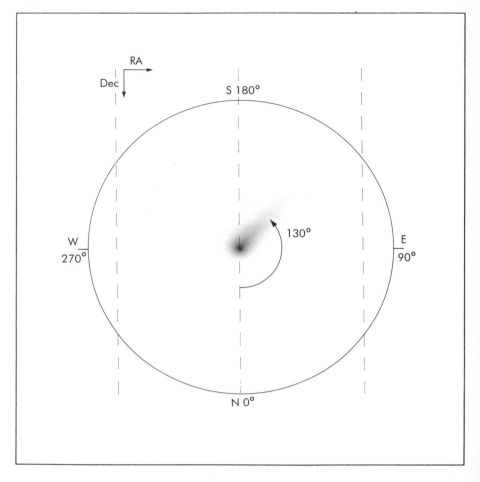

Figure 11.4
Measuring the
position angle of a
comet's tail.

the comet passes through the nodal plane, an apparent-ly Sunward-pointing tail may be seen. The visual obser-vation of comet tails is generally not sufficiently accurate for professional astronomers. Observation is thus best left to photographers and CCD observers. There is space on the observing form to note both the position angle and length of tails, and any asymmetry of the coma. The position angle should be measured near to the head, where the feature first appears; the end of the tail should not be used, unless this is to indicate the curvature. Only record the longest tail in the tabulated data; details of other tails can be included with descrip-tive comments. Figure 11.4 shows the position angle of the tail or coma elongation. North (higher Declination) is 0°, East (greater Right Ascension) is 90°, South is 180°, West is 270°.

How to Improve your Observations

Some simple procedures can help to improve total magnitude estimates, which often differ by several magnitudes between observers.

- Always allow at least 15 minutes to become dark adapted. Not only does the spectral response of your eye change as you dark adapt, but you start to see the fainter parts of the coma, which will make the comet appear brighter.
- Always use the smallest aperture and lowest magnification that will clearly show the comet. Try to use the same telescope and magnification throughout the apparition as this will minimise the systematic effects of aperture and magnification.
- Use several different comparison stars, if possible choosing those that have photoelectrically determined magnitudes and similar colour to the comet (spectral class B, A or F); be careful to identify the stars correctly. Keep each comparison star and the comet in the centre of the field as you make the estimate. Try to use comparison stars which are at the same altitude as the comet. For comets with orbits known well in advance try to obtain BAA variable star charts for fields close to the comet's path.
- Take time over your observation; a hastily plotted, incorrect position can waste a lot of time and lead to an erroneous negative observation. Try to avoid observing when you are physically below standard.
- Do not check the ephemeris magnitude just before you observe – it may be considerably in error.
- Comets often have to be observed in twilight and at low altitude; nevertheless, try to observe with the comet as high and in as dark a sky as possible. Haze, cloud, Moonlight and light pollution will all have adverse effects on accuracy.
- Observe on as many nights as you can, but try to avoid remembering the brightness from the last estimate.
- Observations of comets in the morning sky are particularly useful, as few observers like getting up early!

Submission of Results

Amateur observers are requested to submit their observations promptly to the appropriate organisation, because the results are often used by professionals when planning observing sessions on large telescopes. As much detail as possible should be included in the observer's personal log.

Photography

Photography will continue to be important for tail observations, but developments in CCD imaging are playing an increasingly large part in cometary work.

Photographs give a permanent, objective record of a comet. Negatives or transparencies can be enhanced, either photographically or electronically, to show detail in the coma or tail. The latter aspect is particularly important as tail disconnection events (DEs) give information on the solar wind which is of use to professional astronomers (a DE occurs when the magnetic field carried by the solar wind reverses in direction). As with visual observation, accurate records must be kept, and as solar wind velocities can be very high, times of exposures should be noted to the nearest minute.

For a bright comet, all that is required for a successful photograph is a tripod, fast film and a camera which can take a time exposure. Some of the most spectacular comet photos have been taken in this way. An SLR (single-lens reflex) camera is the most versatile, and one with a 50-mm $f/2$ or faster lens can serve a variety of astronomical purposes. For photography of fainter comets, a well aligned equatorial mount is needed for a guided exposure, either with the camera at prime focus, or mounted piggyback with a telephoto lens. If possible it is preferable to guide on the comet, particularly if it is moving quickly. Some CDD packages can be interfaced to a computer and used to give good guiding, even on a poorly aligned telescope. Film is largely a matter of personal preference, but either monochrome negative, or colour slide, film can give pleasing results. Further details on photography can be obtained from the BAA's Photographic Coordinator or the Instruments and Methods Section.

When submitting photographs it is very important to give the field size (or include a scale bar) and orientation. With transparencies or negatives it is also important to mark which way round they should be viewed.

Astrometric observations, those giving the precise position of a comet, are rather specialised and traditionally require a wide-field telescope such as a Schmidt, and access to a plate measuring machine. However, techniques of CCD observation are now accessible to amateurs and it is possible to carry out astrometry using a CCD and the appropriate computer software instead of a photographic plate. The BAA I and M Section's CCD advisor can offer further advice on CCD astrometry to anyone interested in taking it up. Though the angular field of CCDs is generally too small to obtain high resolution images of tail structure, they are valuable for objectively recording structure in the inner coma, measuring the light distribution in the coma, and measuring the total magnitude. The latter two aspects can also be determined using a photoelectric photometer.

The ST6 camera, using the Texas Instruments TC-241 chip, is red sensitive and needs a V-band filter to obtain useful magnitudes. The Starlight Xpress camera, using the Sony ICX027BL chip, is much closer to the visual band and useful results can be obtained without a filter. Magnitude observations when the air mass is more than 2 are much less reliable. Use photoelectric sequences for the comparison stars, for example the Yale Bright Star catalogue, or underlined stars from the AAVSO atlas. A five-minute exposure may be needed to record the comet, but often only 10–30 seconds for the comparison stars, using separate exposures. Images should be processed in five steps:

- remove dark current
- take mean
- flat field
- clip stars from within the coma
- do photometry (you may need to calculate extinction).

It is possible to make accurate magnitude observations of comets as faint as magnitude 15 using relatively modest equipment. A 400-mm *f*/2 telephoto lens gives a field measuring 1.3° × 1°; this is too short for astrometry, though this can be done with an 800-mm lens. For the total magnitude, include all light within a circle centred on the central condensation, even if the coma is asymmetric. For a very careful reduction the typical error is less than magnitude 0.05 for a magnitude 15 comet and less than magnitude 0.01 for a magnitude 10 comet, the former reducing to magnitude 0.1 when other errors (such as seeing and atmosphere transparency) are taken into account.

CCD observations are now going to magnitude 18 and tie in well with visual observations. Unfiltered CCD total magnitudes are usually within magnitude 0.1–0.3 of the visual magnitude (depending on the chip). However, the use of a V band filter is preferred.

Some faint comets may undergo outbursts. For example, P/Schwassmann–Wachmann 1 seems to be more often in outburst than quiescent. These comets would be good targets for observers with CCD equipment. The CCD can also be used to measure coma profiles, and this may help to resolve the question of consistent DC reporting.

Comet Hunting

To discover a comet you need a lot of perseverance, dark skies and more than a little luck. The average search time for a discovery is around 300 hours, but some people have found a comet in their first hour while others are still searching after more than 1000 hours without success. Either photographic or visual techniques can be used. Sky conditions in the UK are now generally not suitable for successful comet hunting, though there have been some successes in recent years. A systematic approach is needed: first it is necessary to become familiar with what comets look like, as only then will you recognise an interloper. Although comets may appear anywhere in the sky, the best regions to search are those within 90° or so of the Sun, which the Moon has just left; the average discovery takes place with the comet about 25° above the horizon. A common procedure is to search the evening sky as it becomes dark, starting by sweeping along the horizon from North to South, 45° either side of the Sunset point, moving no faster than half a degree per second. The next scan is in the opposite direction, raising the telescope by about half the field of view; this is repeated until you reach an altitude of about 45° above the horizon. With no stops this process will take about two and a half hours. Searching in the morning sky is generally more successful, and is just the reverse of the above procedure.

If you do think you have found a comet, first wait and see if it moves – most suspect objects are still! Beware of clusters of faint stars, which can often appear diffuse under low magnification, and ghost images of bright objects. Once you are reasonably certain then contact the Comet Section of the BAA.

Orbital Computation

Personal computers have taken much of the labour out of orbital computation, and predictions for expected returns of periodic comets are published each year in the ICQ *Handbook*. Predictions and orbital elements for the brighter comets expected to return are published in the BAA *Handbook*. Some software packages enable orbits to be calculated from observations, though as a rule orbits and ephemerides for new comets quickly become available on the Internet.

Calculation of Magnitude

As a comet nears the Sun it loses material into the coma at a greater rate. The rate of loss varies from comet to comet, so predicting the brightness can be difficult. The approximate total visual brightness of a comet can be represented by the equation:

$$m_1 = H1 + 5.0 \times \log(\Delta) + K1 \times \log(r)$$

where H1 and K1 are constants, Δ is the comet's distance from the Earth, and r its distance from the Sun. If the comet was exactly 1 AU from both Earth and Sun, its magnitude would be H1. If few observations have been made it may not be possible to derive a value for K1; in this case a value of 10 is assumed for long period comets and the first constant becomes H10. For short period comets a value of 15 is assumed for K1 and the first constant becomes H15. A purely reflecting body would obey the inverse square law, giving K1 a value of 5.

Historical Research

Belgian amateurs recently made an interesting discovery in old ship's logs of observations of a comet seen off the Cape of Good Hope in 1733. These observations included positions which were good enough to obtain an orbit for the comet. Although the comet was known, no orbit had previously been determined. Many such observations may remain hidden in libraries and museums, and this could be a worthwhile rainy day pastime.

References

Edberg S and Levy D, *Observing Comets, Asteroids and the Zodiacal Light*, Cambridge University Press (1994)

Kronk G, *Comets*, Enslow Press (1984)

Whipple F L, *The Mystery of Comets*, Cambridge University Press (1985)

Yeomans D, *Comets*, John Wiley (1991)

Chapter 12
Occultations

Alan Wells

An occultation occurs when one body passes in front of another and hides it from view. The study of occultations involves the observation and timing of these events.

Types of Occultation

There are two main types of occultation: planetary occultation (including those by the minor planets) and lunar occultation.

Planetary Occultations

Stellar occultations by planets can provide useful information on atmosphere and ring systems if photoelectric recording is applied. Planetary rings have been discovered by recording light from a star being dimmed as it passes behind a ring.

Planetary events can include occultations of a planet's satellites by each other. Jupiter is the most likely planet to have its moons go through a series of occultations with each other. Timing of these events is more for interest than to contribute to any programme of study.

Occultations of moons by their planet often cannot be timed, because the disc of the moon produces a slow fading effect. The problem is further compounded by a planet that has a gaseous atmosphere which will not provide a sharp extinction.

On rare occasions, minor planets occult stars when seen from a particular location on the Earth's surface. Timing of these events requires photoelectric recording if the data obtained is to be useful. The shape of such bodies has been determined by photoelectric timings. Visual observations are possible in a few cases but these require considerable skill.

Lunar Occultations

Lunar occultations are further divided into total occultations and graze occultations. A total occultation involves a single event at a limb. A star can either reappear at a limb (emersion) or disappear there (immersion), the latter being far easier to see as the dark limb is often illuminated by Earthshine.

Graze Occultations

Unlike total occultations, graze occultations are only visible from a particular location and produce a narrow track across the Earth's surface. Graze occultations produce multiple events near the two poles of the Moon. During some total occultations near the poles it is possible to see a star disappear and reappear in a dark region. This type of event is seen if the observer is near to a graze track.

Observation

Predictions

At one time the United States Naval Observatory (USNO) in Washington D.C. was the main source of predictions. These went down to magnitude 12 for observers with larger telescopes. This work is now undertaken by the International Occultation Timing Association (IOTA) and has both American and European groups. IOTA is now the main source of predictions, and is also the main coordinating body for occultation observers.

The IOTA-generated prediction timings are accurate (to within two seconds of time) and also contain other information about the event and its participants, such as the nature of the stars involved and their catalogue numbers.

Following the increased availability of personal computers, other sources of prediction are now possible. Many national groups and other astronomical societies have developed or operate programs which will predict local events. Using this method, predictions accurate to ±5 seconds can be obtained. Unfortunately, in these predictions the stars are not shown as being double or variable.

The accuracy of any predicted timing will depend on the source data and the location of the observer relative to the site of prediction. The location from which an observation is to be made must be known accurately to within 30 metres in all directions. This information can be obtained from local large-scale maps.

Often when a prediction time appears to be displaced by a constant value relative to the observer's site, the resulting error can be judged and a more refined prediction time obtained. It may indicate that the original position given to the prediction program was incorrect.

Timing

Timing is another essential component in recording a good observation, and the choice of method will be influenced by such factors as location and access to telephone or radio time signals. Ideally the accuracy of timing needs to be to a tenth of a second.

Before timing an occultation, it is a good idea to use your timing system against a known time standard. This will let you become familiar with the equipment and show any limitations that might be present. Practise timing any events which are unpredictable, such as the random appearance of dots on computer systems. Most keen computer users will be able to provide you with a suitable program if your local astronomical society cannot. Practise at different times of the day, to build up your experience and improve your reaction time. In some cases many weeks may pass before you get a chance to time a real event, therefore an equipment check is a good idea before you go out to observe.

There are both commercially manufactured and home constructed receivers that enable you to listen to many of the standard radio time beacons. Care has to be exercised with portable receivers as they tend to be directional, and suffer power problems if the wrong types of cells are used at low temperatures. As a rule, alkaline cells are able to work in cool locations, although other types can be employed if they can be kept warm in a pocket.

It is now possible to buy relatively inexpensive clocks that (in most countries) will synchronise themselves with a local time standard transmitter.

Portable telephones are another possible source of time signal, though in some telephone systems a small delay may occur in the arrival of the signal.

Time signals to be avoided include those derived from teletext transmissions, as these are often seconds in error, and verbal checks on radio stations.

First Attempt

Timing of the disappearance of a 6th magnitude or brighter star at the dark limb of the Moon, with the phase below 30%, will give the greatest chance of success.

Method

When a total occultation is to be timed, the following is a typical procedure. A minimum of 30 minutes of preparation time before the event should be allowed, especially when the observer is new to the study.

The prediction should be checked and adjusted to local time or Universal Time (UT). Look under the column headed Ph in the predictions to determine the type of event – either D (disappearance) or R (reappearance). For this example, assume a disappearance at dark limb.

The location of the event on the limb can be found from the CA (cusp angle) in the predictions (see Figure 12.1). The point of occultation is an angle measured from the North or South cusp. A typical CA of 45N indicates the point of disappearance 45° from the North cusp. A minus sign before the CA value indicates that the occultation will occur on the bright limb. These predictions are generally for stars of magnitude 2 or greater.

If CA is not available, the Position Angle (PA) value will be shown. PA is a little harder to use as it is measured from the North point of the Moon through East. As the North point is not so easy to determine exactly, compared with the cusps, it is a good idea to obtain a

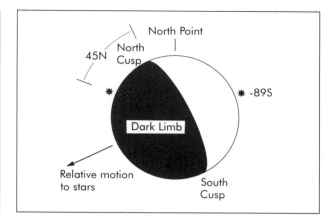

Figure 12.1
Measuring cusp
angle on the Moon's
limb.

prediction which has both. With this a direct comparison between the two can be made observationally, so that the PA value can be located more readily.

During disappearances, when the dark limb is visible in Earthshine, there comes a point when the light from the star cannot be separated from the limb. At this time the star will appear to hang on the limb for 30 seconds or so before being finally immersed.

Looking away from the eyepiece before the event will help to relax the eye and improve the chances of seeing the immersion, though care should be taken not to look away when only a few seconds remain.

The method of timing used will decide the next course of action. Assume that a stopwatch, preferably with a lap facility, and a standard time signal (radio or phone) are used. The stopwatch can be started against a known minute and checked with the lap against subsequent minutes. The watch will then be stopped when the occultation is seen.

An alternative method is to set the stopwatch to zero, calculating the time of occultation by the reduction method described below.

Timing Reduction

1 Start the watch at the moment of immersion.
2 Using the lap timer, stop the watch at the minute marker of a standard time signal. Note the lap value and the time signal value.
3 Stop the watch twice more against successive minutes of the time signal. Average the three 'seconds' components of your lap values.

Example: Timing reduction			
Step 1	Immersion:	Start watch from zero	
Step 2	Lap timer:	Stop #1 = 1m 34.1s	(Time signal = 19h 05m)
Step 3		Stop #2 = 2m 34.2s	
		Stop #3 = 3m 34.0s	(Average seconds = 34.1s)
Step 4	Elapsed time = 1m + 34.1s + 0.3s		= 1m 34.4s
Step 5	Time of event = 19h 05m – 1m 34.4s		= 19h 03m 25.6s

4 Add the result to the 'minutes' component of your first value, then add to this your 'personal equation' time (reaction time: PE. If you are uncertain of your PE, use 0.3 seconds as a standard). The result is the elapsed time since occultation.
5 Subtract the elapsed time from the noted time signal value to arrive at the actual time of the event.

Reporting Observations

The International Lunar Occultation Center (ILOC) in Tokyo is the main centre for data processing, and can provide report forms. Record the timing on the appropriate form as soon as possible. This will reduce the chance of information being lost or wrongly remarked. Check any subtractions that are applied, as errors commonly arise in this situation.

When entering the star identification, use either the USNO or SAO number and the correct identifying letter in the appropriate column. Instructions accompany the report form, which will give further guidance on its completion. It is also possible to submit results on 3.5-inch floppy disk if you have access to a personal computer; the correct data format is available from ILOC and will assist in the transcription process onto their database.

Double Stars

Stars which are close doubles can produce a variety of non-instantaneous events, which include 'fades' and 'step events'. When a double star which is aligned at an angle to the Moon's limb is occulted, the star will appear to alter in magnitude as one component is occulted before the other. Double stars can have very close separations or be aligned nearly parallel to the limb so that they fade during occultation. This take as much as 0.3 seconds, dependant upon the position angle of the stars concerned and the nature of the limb profile at the point of occultation. It is worth checking to see if the star in the predictions is a double, as this will alert you to a possible non-instantaneous event.

Any fades or steps need to be recorded on the back of the report form in the Comments section.

Unpredicted Events

There are also stars which are occulted that do not appear in the predictions. These should still be timed and entered on the recording form. Details of cusp and position angles, together with an estimation in magnitude, will help them to be identified.

Reappearances

Reappearances are the most difficult type of observation to make. The lack of a visible reference point, as well as not being able to estimate the time to occultation, means that for this exercise a more planned approach is required.

The greatest problem when working alone is knowing when the star is due to reappear. If the means to generate an audible time signal are to hand, a shorter period of intense concentration will be necessary before the star becomes visible.

If the CA is known, using a field of about half the Moon's width should ensure success. The fainter the star, the more critical it is that it appears in the centre of the field. If the limb is visible in Earthshine, this reduces the size of the problem.

Good tracking with a reliable motor drive is essential for all but the brightest events. With practice you should be able to visualise the location of the dark limb relative to the bright limb with a particular eyepiece, and this will assist you in locating the field.

It is much better to start with stars brighter than magnitude 5, building up your experience of these events, rather than risking disappointment by beginning with fainter stars.

Preparing for a Graze Occultation

Unlike total occultations, which are lone events, grazes are best tackled by teams if the greatest amount of data is to be collected.

The location of the northern (or southern) limit of the graze will appear in the prediction as a series of map references. This enables a track to be plotted on a large scale map. From this track, suitable sites may be located. As the teams will be at right angles to the track, a road or path in this general line will be ideal. The location in latitude and longitude should also be found for each of the sites.

Typically, the track is about 1 km wide. Too far into the track, and a total occultation will be seen; outside the limit and a near miss is all that can be seen.

Should the prediction information also include a limb profile, the optimum placement of teams across the track of the graze can take place. The profile shows the expected shape of the limb at the point of the graze, together with relative distance from the track limit. The predictions are for sea level, so that sites more than 120 metres above sea level will need to be slightly further South for northern hemisphere observers than the track shows.

Observers with greater experience should be located at the sites where the greatest number of events are expected, and it should be explained that some teams may not see any events at all. The absence of an event can be just as significant as its occurrence in some cases.

A visit to potential sites before the event is essential, as light pollution or obstructions cannot be determined from the map. When inspecting the sites, look out for any overhead cables, as these themselves can be the cause of 'occultations'.

Remember that the police or local residents could be alarmed by the antics of a group of people carrying strange tubes in the early hours of the morning.

When graze occultation timings are attempted, a whole new set of problems occurs. These include the requirement to time multiple events in a short time scale. Some timings may be only 0.5 seconds apart and a few can be just a flash; this will test the observer's ability to the limit. Additionally, a method to record whether the event being timed is a disappearance or a reappearance will be needed.

Where it is possible to do so, use video equipment with time-character generators. This will give a permanent record and allow repeated checking of the event. Regrettably, such systems are only available to a few observers.

Some of the older methods are still valid, such as using a telephone time signal and a tape recorder. Whatever method is used, it is essential to practise before the event and thoroughly field-test any equipment. If a tape recorder is to be used, check on errors within the tape drive. This can be done by recording a time signal for a similar length of time to that required during the graze. If the tape is replayed and checked against a standard time signal, the size of any error should be apparent. Typically, three or more timing sessions should be done, so that an average error can be found. Make the same checks again, moving the recorder around to simulate field conditions. The tape used should be longer in recording time than that required, but not greatly so as this produces extra drag and increases battery drain.

When timing events, a check on the ambient temperature should be made about every 30 minutes. This will be required for entry on the recording form for ILOC.

Pre-event Meeting

A pre-event meeting to confirm who is going where, with which telescope and which timing system, should enable last minute problems to be sorted out. If possible, match an experienced observer with a novice and do use pairs.

During the occultation, many events may take place, so the timing method using a tape recorder and an audio time signal are ideal. The observer then calls 'in' and 'out', or 'flash', which is recorded on the tape along with the audio time signal. The reduction of data to a time can be checked more easily if a common system of words is used. If a multiple-event stopwatch is used, the second member of the team will need to record the type of event being seen.

As soon as possible after the event, timings should be gathered together and checked, along with any other relevant information collected by the teams. Whichever method has been used, remember to keep a copy of the report in case the original is lost.

It should be clear by now that a great deal of planning

and liaison has to be undertaken by the team leader of graze-observing expeditions. If you have not done this before, talk to someone who has, build on their experience and, preferably, invite them along.

Equipment

The type of telescope employed and its location in a observatory or open site can have a considerable effect on the success rate. Observatory instruments often have a limited field of vision, which portable instruments do not. Portable instruments that can be tracked and set up with shelter from the wind are the most useful type.

Tracking of the telescope must be good, to retain the star image in the centre of the field and reduce the distraction of constant correction.

It is a good idea to have both types of instrument available, to meet the full range of observation opportunities. It is considered that a 6-inch reflector or equivalent is the preferred minimum, but timings can be obtained using smaller telescopes. The number of events which can be observed is a function of telescope size and stellar magnitude. This is easy to understand if you think of how many stars there are for any given magnitude. Larger telescopes are able to see many more of the fainter stars, so the greater the aperture, the more occultations can be seen.

The apparent relationship between magnitude and aperture can be upset if the percentage of the bright limb is greater than 80%, or cloud conditions are limiting the visibility of fainter stars. Generally stars brighter than 8th magnitude are the most commonly timed, because this can be carried out under less than perfect conditions during the timing period.

With high percentages of bright limb illumination, high magnification can further improve things by reducing that part of the limb that may be seen in the field. This technique is limited when the illumination is greater that 80%, as the star then has to be located on the edge of the field. One method of overcoming this problem is to use an occulting shield in the eyepiece. This does not have to be an expensive device and can be made with something as simple as a plastic bottle top, cut to section with a craft knife. A range of such occulting shields with different sections will provide the optimum field for any given situation. If the eyepiece is rotated with the shield in place, the best result can be judged.

An alternative to an occulting shield, which is opaque, is to use a section of neutral density filter. The big advantage of this method is that the location of the edge of the dark limb can be better estimated, thus allowing a better guess at when the star will be occulted.

The size of the field of a given eyepiece relative to the Moon's disc is an essential piece of information to an occultation observer. By experimenting with the full Moon, and a range of different types of eyepieces, the location of the dark limb can be better estimated during subsequent observations. When testing eyepieces for field size with undriven telescopes, time how long it takes for a star to cross the field at its widest point. Knowing this time will allow you to move the telescope, so that the star will be centred in the field at occultation.

Care should be taken to ensure that the reflective light scatter from the inner surfaces of the instrument, particularly the eyepiece tube, is low. Surfaces are generally matt black, but this can become very reflective if high humidity levels exist. Lining with irregular surfaced materials will improve things. Matt black cotton fabric bonded to these surfaces will absorb a lot of moisture without producing reflective effects. Similarly, black paper can be used, though this can become a little more reflective. In open tube Newtonian equipment, secondary reflections can be set up on the surface adjacent to the eyepiece hole which are then picked up by the secondary mirror and integrated with the main image. More anti-reflective material around this area will improve the situation. The problem can be further aggravated by internal reflections within the eyepiece elements. Care must be taken to choose types which suffer least from this problem, which is known as ghosting.

Trial and error will show the best location of the instrument for a given site. Particular care is needed to locate a position where a sudden light from a neighbour's property will not be a problem. Interest your neighbours in the work if possible, or explain the problems you can have; most are willing to cooperate.

Factors Affecting Success

The chance of obtaining a successful timing of an occultation is dependent upon a range of factors, many of which are outside the control of the observer. Among these problems is the weather; cloud and the general state of the atmosphere are the main limiting factors in this type of study. Perhaps the largest cause of failure is very thin cloud, which is almost invisible to the naked eye but causes scatter of light from the Moon's bright limb into the area of the dark limb where the event is to occur. Unfortunately, thin cloud is not the only problem likely to be encountered. Mist can produce similar effects, as can dewing on secondary mirrors.

The observer's breath condensing on cold optical surfaces also causes problems. Try not to breathe on the eyepiece, instead exhaling in a direction away from it.

Dark skies, a pre-requisite for some types of astronomical observation, are generally not such a major requirement for observing occultations, making timings from city areas possible. It is also possible to see occultations of stars brighter than magnitude 3 during the day.

With all aspects of this work, practice and the assessment of prevailing conditions can affect the outcome in a dramatic way.

I very much hope that you will try to time some of the events described, contributing your share of data to the scientific world.

Reference

BAA Handbook for Lunar Observers, British Astronomical Association (1992)

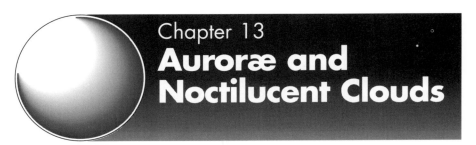

Chapter 13
Auroræ and Noctilucent Clouds

David Gavine

The Aurora

Of all natural phenomena, the Aurora Borealis, with its southern and simultaneous counterpart the Aurora Australis, is one of the most beautiful. Briefly, it is caused by interactions between the magnetic field of the Earth and that of the solar wind whose electrons, and some protons, are accelerated to energies of several thousand electron volts and precipitated into the upper atmosphere. They excite the atoms and molecules of rarefied oxygen and nitrogen, and the various electronic transitions emit light of certain wavelengths which give the aurora its distinctive colours, although there is also a strong ultra-violet emission which is observed by satellites. To Earth-bound observers the aurora is visible from about 100 to several hundred kilometres above the surface. Radio amateurs detect distinctive auroral phenomena such as 'ghostly' Morse and voice transmissions, and the accompanying disturbances in the Earth's magnetic field can be monitored with simple apparatus which can be constructed by the amateur.

There is a more or less permanent oval of excited atmosphere around each magnetic pole. In quiet conditions these rings of light remain fixed in space relative to the Earth–Sun line and the Earth rotates underneath, so that each day at 'magnetic midnight' (around 2200 UT in Britain) the observer is carried round to the point where the oval is at its greatest equatorward extent. During enhanced solar wind activity the Earth's magnetic field becomes disturbed and the auroral oval brightens on the night side, expands, and sometimes breaks up. Auroral light may then be seen at lower latitudes. Satellites may also detect a 'bridge' of quiescent light inside the oval itself, partly or wholly crossing it along the noon–midnight line. This discrete polar-cap or 'theta' aurora can be seen only in the 24-hour darkness of winter in very high latitudes and has recently been investigated by intrepid amateur astronomers in Spitzbergen.

Studies over many decades show that there is a zone passing through the north of Norway, Iceland, Labrador, Hudson Bay and Alaska where auroræ are seen most frequently, that is, almost every clear night. Because the north magnetic pole lies in northern Canada, observers at any given latitude in North America enjoy more auroral displays than those at the same latitude in Europe or Asia. In the southern hemisphere there are, of course, fewer observers, but auroræ are seen quite often in the South Island of New Zealand, and occasionally in southern parts of Australia.

Magnetic substorms are small, frequent and short-lived changes in the interplanetary magnetic field (IMF) of the solar wind near the Earth. The onset is rapid, and the changes in the auroral oval may include a westward surge of wavy structures from the midnight sector of the sky into the dusk sector. Auroræ are then seen in Scotland, Scandinavia, Canada and the northern States of the USA. However, bigger disturbances, associated with major solar flares, usually (but not inevitably) produce more dramatic events which can last for several days and push auroral light further equatorwards. One of the most memorable of these was the great storm of March 13/14 1989 which gave brilliant auroral light over the entire sky in the south of England.

The aurora is not entirely absent at solar minimum. Often, quite diffuse auroral light, sometimes a few rays, can be seen in the north of Scotland or Ireland. These are the result of coronal holes on the solar surface, visible only in X-rays, from which electrons spray out into the solar wind. The auroræ they produce are quieter and longer-lasting than those of the transient events, so that a persistent active area on the Sun may produce repeating visible or radio auroral performances every solar rotation, and these show up on 27-day Bartels diagrams. Coronal holes reach a maximum around the end of one sunspot cycle and the beginning of the next. Although storm-type auroræ are much reduced in frequency around minimum they can still happen unexpectedly:

one of the biggest auroral storms of the century, on February 8/9 1986, covered the skies of southern England in blood-red light and was visible even in Hawaii.

Where and When?

It is not quite true to say that the further north you go, the more auroræ you see. Frequency falls off in the 'trans-auroral' zone close to the geographical pole, and there is certainly no need to go to the Arctic, where winter temperatures make observing very uncomfortable, batteries and camera mechanisms fail and film becomes brittle and cracks. Northern Norway, Iceland and Shetland are best avoided too, for wet, overcast conditions may persist there for weeks, at any time of year. Then, north of about latitude 55°, it does not get sufficiently dark in midsummer. The experience of many observers over the last four solar cycles suggests that the best places to see the aurora in Europe are Finland, the island of Bornholme and north-east Scotland, especially around the Moray Firth. North Americans are more fortunate: the southerly dip of iso-auroral mappings means that auroræ are frequently seen not only in all parts of Canada but in the States of Minnesota, Montana and North Dakota in the USA. The well-known observer Jay Brausch, of Glen Ullin, North Dakota, (at geographical latitude 47°N – which is at about the same geomagnetic latitude as Fair Isle) is favoured with clear skies and has observed and photographed over a thousand auroræ in ten years.

It was once thought that transient auroral activity peaked with the sunspot maximum. Again, that is not true: the Sun may be covered in spots and the aurora absent, and vice versa. Experience has shown that most storm and substorm auroræ occur around the build-up to sunspot peak, a year or two before it, as shown by the high incidence of reports in 1978 and 1989, and there tends also to be quite a lot of auroral activity about a year after sunspot maxima. In any one year, too, auroræ are likely to occur around the equinoxes.

The aurora is an unpredictable phenomenon for which long-term warnings are seldom reliable. On occasion the media have alerted the public to huge displays on certain nights – and nothing happened. The author has been caught out by complacency, once telling an audience that things were quiet on the auroral scene and

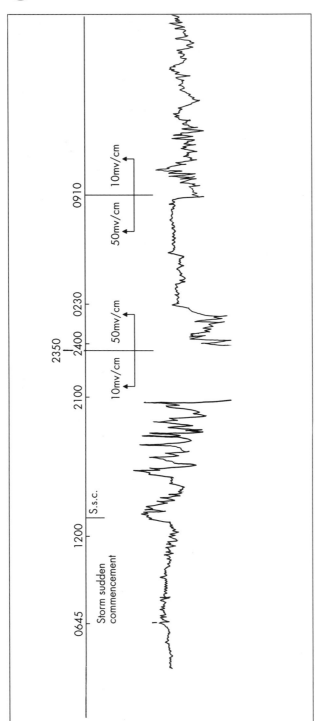

Figure 13.1 A magnetic storm, recorded by a Fluxgate magnetometer.

witnessing an all-sky storm the same night. There are short-term clues: monitor the Sun, watching for the break-up of big spots or spot-groups near the central meridian. If you have a simple magnetometer, watch out for the tell-tale pattern of the approaching storm: a 'blip' signals the compression of the Earth's magnetic field by a solar wind disturbance. A slow rotation of the magnet westwards followed by a more rapid movement eastwards is often the warning sign. Remember, however, that these are not always the harbingers of a visible aurora – depending on the polarity of the solar wind the auroral oval may expand northwards into the 'trans-auroral' regions so that observers south of the oval see nothing. The literature contains instructions for building and using magnetometers to detect these tiny fluctuations, from the simple 'jam-jar' type to sophisticated electronic devices, and experience will soon teach the observer to distinguish between magnetic storms, small regular diurnal variations and local disturbances such as trains. Figure 13.1 shows the trace of a storm produced by a Fluxgate magnetometer. Radio observers sometimes detect auroral interference in the afternoon which could manifest into a visual display at night, but radio auroræ may occur strongly in the absence of any visual display and vice versa. Amateur journals such as *Radio Communication* (Radio Society of Great Britain) give probabilities of ionospheric activity based on Bartels diagrams, but these are for recurrent events.

Serious auroral observing is best done by a group – say, in an astronomical society – at least one member of which has access to a magnetometer, and one to a clear northern outlook untroubled by light pollution. At the first signs of something interesting on the meter, the others are alerted by telephone. Check the sky every clear night, not just at magnetic midnight, because a storm-type aurora can be well under way by early evening.

Observing the Aurora

A typical auroral storm usually begins with the appearance of a glow on the northern horizon followed by the development of an arc which rises above the horizon (see Figure 13.2). The arc may then become narrow and bright, rise higher, split, or develop bright blobs which break into short vertical rays, at the same time folding into a 'band', the 'wavy structure' of the westward surge

(see Figure 13.3). Movement East or West, for example, may be seen of the rays themselves or of kinks along the base of the band, and there may be fluctuations in brightness. As the rays brighten and lengthen the distinctive green colour (the oxygen emission, 557.7 nm) becomes apparent and the rays or ray-tops may show red (oxygen 630.0 nm). The rays may converge to a point nearly overhead, the magnetic zenith to which the dip-needle points, to form a corona (not to be confused with coronal holes on the Sun), while auroral light may spread into the southern sky as a pale green or red veil. The aurora may then break up into diffuse pulsating patches which gradually fade, and the whole performance may repeat about an hour later, and several times during the night. Most auroræ in mid-latitudes go no further than the glow or quiet arc stage which can last for hours.

Figure 13.2 An arc, an early stage in the development of a typical aurora.

It is not necessary for the observer to write a description of every auroral form in the sky, or to give a minute-by-minute account, but it is important to note the times of major events such as changes of pattern, e.g. the break-up of an arc into rays, sudden brightenings, fadings, onset of movement, coloration and coronal peaks. It is useful for the observer to make up some kind of simple form to record the observation, and simple sketches help. The record should include the following details.

Session Details
The observer's name; latitude and longitude; date

Figure 13.3 Rays and wavy bands, a further stage of auroral development.

(always in the form of the double-date, e.g. 1994 April 12/13) and time (always in Universal Time (UT)).

Measurements

Try to measure the altitude of the base of an arc with a simple home-made theodolite or alidade, to the nearest degree. This enables the arc to be plotted on a map of geomagnetic latitude. Angles in the sky can be estimated roughly – at arm's length the spread of the fingers is about 10°, although of course this varies with individuals. If instruments are not used, altitudes are invariably over-estimated. Measure also the maximum altitude of the auroral light above the northern horizon – in a big aurora this can be more than 90° if the light extends beyond the zenith. Altitude of base of arc or band is denoted by 'h', and of the top of the display by 'l'.

Condition

Auroræ are quiet (Q), or active (a, p) with movements or variations in brightness:

a_1 Folding of bands
a_2 Rapid change of shape of lower border
a_3 Rapid horizontal movement of rays
a_4 Fading of forms with rapid replacement by other forms

p_1 Slow pulsations
p_2 Flaming: this spectacular phenomenon is often seen

Table 13.1. Notation used to record an observed aurora

Condition	Qualifier	Structure	Form	Brightness	Colour
Q	m	H	A	1	a
a_{1234}	f	S	B	2	b
p_{1234}	c	R_{123}	P	3	c
			V	4	d
			R		e
			N		f

near the peak of a big display. Waves of light seem to rush from the horizon to the zenith, lighting up forms. It looks rather like a wind blowing over a field of wheat

p_3 Flickering: rapid changes of brightness, seldom seen outside the auroral zones

p_4 Streaming: irregular horizontal variations of brightness in homogeneous forms

Qualifying Symbols
c Coronal – the rays or patches converge overhead
f Fragmentary, e.g. partly-formed arc
m Multiple, e.g. m_3R_1A = three rayed arcs

Structure
H Homogeneous, lacking in structure
S Striated, with fine horizontal filaments often seen at high elevations
R Rayed. Lengths of rays are denoted R_1 (short, 10–20°), R_2 (30–50°), R_3 (60°+)
Thus, R_1A is a rayed arc with short rays, R_3B is a rayed band with long rays. R_3R means a long ray on its own

Form
N denotes a horizon glow or light in cloud or other doubtful structure
V Veil, a faint light, usually as a background
A Arc, a rainbow-shaped structure
B Band, like an arc but with one or more kinks or folds
R Ray, like a vertical searchlight beam
P Patch or surface, a large blob of light

Brightness
1 Faint, like the Milky Way
2 Comparable to moonlit cirrus. Colour detected
3 Comparable to moonlit cumulus. Colour obvious

4 Much brighter than 3, casts shadows, seldom seen outside auroral zones except in big storms

Colour
a Red in upper part
b Red lower border. This is a nitrogen emission caused by very energetic electrons reaching lower levels
c White or green
d Red
e Red and green (atomic oxygen) sometimes distributed horizontally e.g. red and green rays along a band
f Blue or purple dominant. Blue ray tops are sometimes seen when very long rays are sunlit

This aurora 'shorthand' is written in the order shown in Table 13.1. For example:

QHA1c = faint quiet white homogeneous arc;
$p_2m_3R_3B3e$ = three bright flaming rayed bands with long red and green rays.

However, handy though it is, many observers do not bother to use the code. A clear, honest description in simple words, with a sketch or two, is just as good. The illustrations in Figure 13.4 show the major forms and variations.

Colour

A word here about colour. Some members of the BAA's Aurora Section view through narrow pass-band filters which transmit the green 557.7 nm emission and block out light pollution and moonlight. Unfortunately these are very costly and almost unobtainable now. As a cheap substitute, use a sheet of green celluloid or Perspex to enhance the green auroral light. Observers differ greatly in their ability to detect red auroral emissions at low light levels, and a red filter can help by suppressing other wavelengths.

Photography

Photography of the aurora can be done with simple 35-mm cameras of focal length $f/2.8$ or faster. Good films for the purpose are Kodak Ektachrome 400 and Fuji 400, both of which are sensitive to auroral red and green emissions. Faster emulsions are also worth a try.

RA Rayed arc

RB Rayed band
(R_1A, R_2A, R_3A, R_1B etc. depending on the length of the rays)

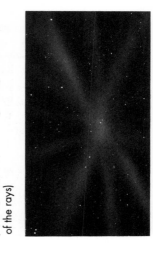

CRR Coronal rays - Converge to a point at the magnetic zenith

HP Homogeneous patch

RR Rays, single or in bundles
(R_1R, R_2R, R_3R according to length)

N Horizon glow or unidentifiable auroral structure

HA Homogeneous arc - a structureless arch of light

HB Homogeneous band - like an arc but with kinks or folds

Use a rigid tripod and a cable-release, and wait for the aurora to brighten and remain fairly static. The trick is to match the shortest possible exposure to the movement, to catch enough light and the sharpness of the forms. Bright aurorae can be photographed in 10–20 seconds, with the bigger the aperture the better – $f/1.4$ is ideal. For a faint quiet arc, a minute or two with a wide angle may be needed. Always record the time and details of each exposure.

Why?

Why do amateurs observe the aurora? The magnetic field is constantly monitored by observatories and the aurora itself can be studied by satellites and all manner of remote-sensing devices, so the importance of the amateur observer is much less than it used to be. However, there is still a small niche. Professional scientists tend to work in the Auroral Zones, but amateurs such as the Aurora Section of the British Astronomical Association compile data banks of sightings world-wide to maintain the continuity of many years of work. These show, for example, how far the aurora extends equatorwards. There is also some interest in the so-called 'flash' aurora, a little-known phenomenon in which brief bursts of light, perhaps only a few seconds in duration, appear in low latitudes. Amateur aurora data are often requested from the BAA by government and academic research bodies. However, to the lover of astronomy, the aurora is well worth watching just for its own sake!

Noctilucent Cloud

It is important to observe this phenomenon, because the compilation of data and statistics on Noctilucent Cloud (NLC) is almost wholly in the hands of amateurs, and the more observers there are, distributed widely in latitude and longitude, the less chance there is of an occurrence being missed.

These beautiful ice-crystal clouds of the Mesophere, at about 83 km altitude, appear only at geographical latitudes from about 50° to 70°, and in the northern hemisphere are seen only between about the last week of May and the second week of August. Below 50° they do not occur, above 70° the summer sky is too bright to see

Figure 13.4 (opposite) The major forms and variations of aurorae.

Figure 13.5
Bands of noctilucent
cloud in the sky over
Edinburgh.

them. Eastern England and Scotland are good NLC-watching localities, as is Denmark. Figure 13.5 shows bands of NLC photographed in the sky above Edinburgh.

The clouds appear in the twilight arch when the Sun is somewhere between 6° and 16° below the horizon, and they shine with a pearly white, blue or golden radiance. Their fine structure can be studied in binoculars. They have a superficial resemblance to cirrus or cirrocumulus, but even high tropospheric clouds are in darkness when the NLC appears, and they are seen silhouetted against it. Beware of twilit or moonlit cirrus near the zenith! If there is no obvious NLC towards the north, what you see is liable to be spurious. Beginners are easily confused.

There are five basic structures to note. These are:

I   Veil: a structureless sheet, usually seen as a background to other forms.
II Bands: straight or slightly curved lines, nearly parallel or crossing at small angles.
III Billows: large or small wave-like structures like herring-bone, ripples on a sandy beach or altocumulus cloud.
IV Whirls: complete or partial rings or loops.
V Patches: small discrete units, brighter than I.

They can be graded for brightness according to the following index:

1 Very faint, just visible to the naked eye or in binoculars.
2 Moderate.
3 Brilliant and intense.

NLC do not move around like the aurora. Make observations four times every hour, on the quarters. Measure with an alidade the topmost elevation of the display above the horizon, and give the azimuths (true bearings) of the east and west limits. Note also the azimuth of the brightest part of the display.

Photography is most important for research. Almost any type of colour film will do; Kodacolor Gold is very good. Alternatively, use high-contrast monochrome film such as Ilford FP4 (125 ISO). Use a standard 50-mm lens and, as a general rule, expose 200 ISO film for 5 seconds at $f/2.8$ for an average NLC. All photographs should be taken from a fixed bracket pointing north, and timed precisely on the quarter-hours. If stars appear on the photographs they will be useful for parallactic height measurements. Remember to use double-dates, UT, and to give the exact location.

Auroræ and NLC were once thought to be incompatible; that is, the electron influx heated the mesosphere, thus preventing NLC from forming, while NLC tended to be seen on a greater number of nights around the solar minimum. However, some recent reports, notably from Canada, indicated that both phenomena might be seen in the sky simultaneously. If you see this, remember that it is rare and of the utmost interest to aeronomists. Record carefully the times of the start and end of such an occurrence, and time any changes such as fades or enhancements of one or the other. Report all of your observations to the coordinators of NLC or aurora in your country, or to the Director of the BAA's Aurora Section, who has links with observers in the rest of the world.

Further Reading

Solar Activity

Baxter W M, *The Sun and the Amateur Astronomer*, David & Charles, Newton Abbott, Devon (1973)

Hargreaves J K, *The Solar–Terrestrial Environment*, Cambridge University Press (1992)

Phillips K J H, *Guide to the Sun*, Cambridge University Press (1992)

Magnetism

Flodqvist G, 'Detecting the Polar Lights', *Sky & Telescope* 86:4 (1993), 85–87

Livesey R J, 'A "Jamjar" Magnetometer', *J. Brit. Astron. Assoc.* 93 (1982), 17–19

Livesey R J, 'A jamjar magnetometer as "aurora detector"', *Sky & Telescope* 78 (1989), 426–432

Pettit D O, 'A Fluxgate Magnetometer,' *J. Brit. Astron. Assoc.* 94 (1984), 55–61

Smillie D J, 'Magnetic and radio detection of auroræ', *J. Brit. Astron. Assoc.* 102 (1992), 16–20

Auroræ

Akasofu S-I, 'Aurora Borealis', *Alaska Geographic* 6:2 (1979)

Bone N, *The Aurora, Sun–Earth Interactions*, Ellis Horwood, Chichester (1991)

Eather R A, *Majestic Lights: The Aurora in Science, history and the arts*, American Geophysical Union, Washington (1980)

Davis N, *The Aurora Watcher's Handbook*, University of Alaska Press (1992)

Noctilucent Clouds

Bone N, 'What are Noctilucent Clouds?' *1993 Yearbook of Astronomy*, ed. Moore P, Sidgwick & Jackson, London (1992), 223–234.

Gadsden M and Schröder W, *Noctilucent Clouds*, Springer-Verlag, London (1989)

Chapter 14

Variable Stars

Melvyn D. Taylor

Our Milky Way Galaxy contains an estimated 10^{11} stars in a variety of evolutionary stages and in two major zones or populations. Those stars generally in elliptical orbits high above the galactic plane are termed Population II stars, while Population I types lie in the plane of the Galaxy. Stars showing a measurable change of brightness are classed as variable stars; these provide clues to the nature of stars which range from hypergiants to white dwarfs. Their study may help the understanding of the Solar System's formation, and perhaps of life itself. For cosmologists, variable stars provide distance indicators which are fundamental to measurements of the Universe. Astronomers for the last few centuries have observed and classified these stars using a host of observational techniques and analytical methods. With so many galactic stars available, some 45,000 catalogued variables (including stars suspected of variability) and only several hundred variable star specialists world-wide, it is obvious there is much work to do. In recent years links between professional and amateur astronomers have brought about a rapid exchange of information (e.g. the discovery of Nova Cygni in 1992), resulting in new findings about stellar phenomena.

From the point of view of the amateur, the basic observation in the study of variable stars is the estimation or measurement of a star's brightness (magnitude) at a particular time. The variable is compared with stars of known magnitude whose brightness is invariable, and the deduced magnitude is plotted as a graph against time to form a light-curve.

Classification

Classification of variables as recognised by the *General Catalogue of Variable Stars* (fourth edition) is based mainly on optical characteristics, and comprises six main categories. These are: eruptives, pulsating, rotating, cataclysmic, eclipsing and X-ray sources; within these classes there are about fifty types and several subtypes. In the original naming of variable stars, some were given Greek letters (Mu Cephei, Beta Persei), others Latin letters, then a systematic lettering system beginning with R (e.g. R Andromedæ). The genitive of the name of the star's constellation forms the concluding part of its name. After 334 are named in the lettering system, the next is designated V335, followed by the constellation derivative.

Eruptive variable stars, as the name implies, have turbulent changes in their 'atmospheres' which show up as changes in brightness. Star types in this class include the well known prototypes FU Orionis (FU), Gamma Cassiopeiæ (GCAS), T Tau (INT), R Coronæ Borealis (RCB) RS Canum Venaticorum (RS), P Cyg (SDOR) and UV Ceti (UV).

Pulsating variables have 'atmospheres' which periodically change, either in a uniform or a non-uniform manner. This major class has many types: Cepheids (CEP), W Virginis (CW) stars, Delta Cepheids (DCEP), irregular variables (L, LB, LC), Mira stars (M), RR Lyræ (RR), RV Tauri (RV) and semiregular variables (SR, SRA, SRB, SRC, SRD) representing a large proportion of objects. There is an enormous range of star sizes in this class; for example, the Alpha Cygni (ACYG) supergiants may have diameters sixty times that of the Sun, but the ZZ Cet (ZZ) pulsating white dwarfs are only about 0.02 times the Sun's diameter.

The rotating variables show variations which are due to the axial rotation of an ellipsoidal stellar object, or of 'star-spots' rotating with the star. The Sun is regarded as a BY Draconis (BY) variable, a type characterised by showing small (0.01 to 0.5 magnitude) visual amplitudes over intervals varying from a fraction of a day up to about 120 days. Alpha^2 Canum Venaticorum (ACV) stars are main sequence objects with rotation periods of between half a day and 160 days; the brightness, the magnetic field and the spectrum all change over this time interval. However, the visual brightness changes are in the range of 0.01 to 0.1 magnitude, a difference which is beyond detection by the visual observer.

The cataclysmic variables (CVs) represent a class of variable objects showing major outbursts of energy at many wavelengths, which may be repeated. Supernovæ (SN) have outbursts of 20 magnitudes or more which may be particularly catastrophic! The novæ (N), recurrent novæ (NR), dwarf novæ (UG), polars (AM Herculis type) and 'symbiotics' (ZAND) are the other types of star in this class. The CVs are interacting binaries, with matter being transferred from a large cool star to a smaller white dwarf via a 'gravitational well' in the system, and a few show an eclipse of one star by the other, e.g. HT Cas, Z Cha and IP Peg. Eclipsing binaries (E) form about 65% of those stars catalogued as variables, and it is estimated that between a quarter and a half of all the stars in our Galaxy are binary in nature. There are three major types, represented by Algol (Beta Persei), Beta Lyræ and W Ursæ Majoris, with many sub-types based on a classification of the light-curve, the degree of interaction of the two stars and the stellar characteristics.

X-ray sources are mostly close binaries in which one star may be a white dwarf and the other a cooler object, producing rapid visual changes of up to 9 magnitudes and fluctuating or pulsed X-ray emission. V616 Monocerotis, which brightened to magnitude 11 in 1917 and magnitude 10.4 in 1975 (when it was bright in X-rays), is classed as type XND, an X-ray nova-like object.

Other types of important variable objects encountered in the *Catalogue* are: BL Lacertæ star-like galaxy nuclei (BLLAC), variable galaxy nuclei (GAL), and quasars or variable quasi-stellar objects (QSO).

The interrelation of several well-known variable star types with giant, main sequence and sub-dwarf stars in our galaxy is shown on the Hertzsprung–Russell diagram, Figure 14.1.

Charts

Variable star organisations offer advice, information and publications to the observer and researcher. Charts available from these groups cover many types of stars and will show: the variable's designation and name, its position in Right Ascension and Declination, a scale chart of the sky with the variable marked as a circled dot (or circle) among field stars, and a set of comparison stars together with their magnitudes. It is important

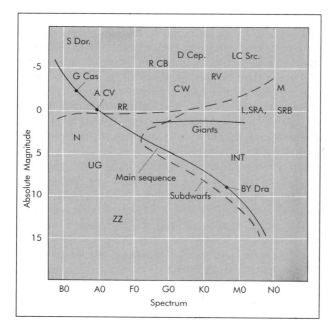

Figure 14.1 Some classes of variable star, shown on the Hertzsprung–Russell diagram.

Solid line:
    Population I stars
Dashed line:
    Population II stars

that the observer uses only those charts issued by the organisation he or she belongs to; another group may give comparison star magnitudes with slightly differing values. A large-scale star atlas is a wise acquisition for use in conjunction with the charts.

Observation

Instruments

Using binoculars and two eyes is more relaxing than peering into the eyepiece of a telescope. It is a more convenient means of observing than opening the observatory/shelter and 'setting up' a large, mounted telescope. A 10 × 50 (50-mm aperture) pair of binoculars will allow numerous variables, to about magnitude 9, to be followed. Magnifications over ×10 require the binoculars to be supported rigidly on a tripod or a custom-made observing chair; a mounted instrument will enable the observer to detect objects fainter than will a hand-held (braced) one.

A typical 200-mm reflecting telescope with a magnification of ×80 may detect magnitude 14 stars, and some observers with 440-mm reflectors have reached magni-

tude 16 and slightly fainter. The limiting magnitude of
any optical instrument depends on several factors,
which include aperture, focal length, magnification,
optical design, sturdiness of mounting, sky brightness
and even the mood of the observer! The long-used
method of setting the Right Ascension and Declination
circles on an equatorially mounted telescope, to find
faint or different variables, has competition today from
computer guidance controls and indicators which will
fit most forms of mounted instrument.

The use of a good eyepiece, compatible with the
quality of the objective lens, cannot be stressed enough.
A higher magnification and a smaller, darker field of
view, may 'pull out' a variable near its minimum brightness. A black hood covering the observer's head and the
area around the eyepiece, thus cutting out extraneous
light, helps to reach fainter stars, as does the obvious
necessity of good eye health.

Light Estimates

The human eye is a remarkable natural photometer,
with a sensitivity which ranges from about 400 to 700
nanometers – from blue to red – and, when adapted to
darkness, the sensitivity peaks at around 530 nm. Given
optimum circumstances, the accuracy of a single
observer's estimate may be ±0.2 magnitudes, to a limiting magnitude of about 16.5. Observers should make
themselves aware of the sources of error inherent in
visual work. Very general advice about attaining good
quality light-estimates would involve a practice of being
careful, consistent and systematic in the methods used.
Errors caused by not allowing the eye(s) to adapt under
dark conditions, a process which may take 10 to 15 minutes, are considerable. Dark adaptation when referring
to charts and making notes is maintained by using a
faint, red-coloured light.

The accuracy of a light estimate is indicated by a
class code given to the observation when it was made. A
class 1 estimate is a reliable (±0.1 magnitude) observation. Class 3 is a poor observation made with the Moon
nearby, or with haze, and is just about worth recording;
a class 2 estimate has a ±0.2 magnitude accuracy.

The timing of an observation is normally given to the
nearest minute in Greenwich Mean Astronomical Time,
which is directly related to the Julian Date method of
time-keeping; GMAT is Universal Time minus 12 hours.

Records of the light estimate are vital, and most observers develop a personal log. An extract from a typical observational notebook is reproduced in Table 14.1.

If the observations are to be sent to an organisation, a report is made from this log on to a standard sheet.

As well as making good quality light estimates, the observer should try to follow a star for as long as possible, taking every opportunity to use a clear night sky. Rapid verification and announcement of unusual or important stellar variability, e.g. the discovery of a nova, is increasingly being made by electronic methods of communication.

The pooled work from several observers on a particular star should provide a continuous light-curve, but the proximity of the Sun has to be considered.

Stars with amplitudes of one magnitude or more are available to the visual observer who uses the naked eye, binoculars or a telescope. The learning and training involved, using visual methods of scaling, recognition and identification of stars and other celestial objects, will be advantageous for those who progress to fainter objects. The committed observer may take on the task of following 60 to 100 variables, of a variety of classes and types, but the 'average' enthusiast will follow a smaller programme. The rare specialist may concentrate on one type of variable, choosing, for example, from the M, SR, RR, N, NR, SN, UG, ZAND and E stars.

Table 14.1. Example of an observing log of variable stars

Variable Des.	Const.	GMAT h – m	JD + dec 2449...	Light estimate	Mag.	Class	Notes	Date yr/m/dd
BN	Gem	07–38	415.32	C (3) V (2) F	6.5	2	8x40B	1994
DW	Gem	07–50		N–3 P+4	9.3	1	R200x44	Mar 3/4
U	Gem	07–53	.33	[P	[12.9	1	R200x44	
IS	Gem	07–55		D (1) V (3) F, =E	6.2	2	8x40B	
AQ	And	08–10	.34	F (3) V (2) K, =H	8.5	1	16x0B	
γ	Cas	08–12		=C	2.3:	3	NE Haze	
RV	Mon	08.30	.35	F (3) V (1) G	7.9	1	16x70B	
SX	Mon	08–31	.35	G–1 H+2	8.1	2	16x70B	

Note: Columns headed *Julian Date + decimal* and *Magnitude* are completed after the observations have been made

Observers with a limited amount of observing time may find the pulsating types, L, M, RV and SR, to be of interest, since these only need to be observed every few days. Most of the cataclysmic variables deserve observation several times a night when in outburst, and every night until they are lost near the limit of the telescope. The novæ and supernovæ come into this category. The NR, NL, UG, ZAND and AM stars, especially the fainter ones, offer original programmes for the owners of larger telescopes. Nova and supernova discoveries, super-maxima of UGSU types, and outbursts of recurrent novæ (NR), should be reported promptly to a variable star organisation, which will, on confirmation, pass the news on to other observers.

The aim of following eclipsing binaries is to time the minima for determination of the system's period. If a change or correction to the orbital period is found, this is significant information which needs to be reported. The observer needs predictions for the times of minima, and the appropriate organisations will supply these. Similar predictions are useful for the maxima times of Cepheids and RR Lyræ pulsating variables.

Table 14.2 gives a guide to the frequency of observation of some variables.

Visual estimates, wherever possible, should involve two comparison stars, one brighter and the other fainter than the variable. The fractional method, e.g. A (2) V (3) B, is a direct comparison between three stars; A is the brighter, B the fainter comparison, and V the variable. In this example the variable is judged to be 2/5 of the difference in brightness between A and B. This is the best method for use by beginners.

The Pogson step and Argelander step methods use independent comparison stars, but it is good practice to use another comparison as a check. Pogson's method requires the observer to detect a difference measured in tenths of a magnitude. A light estimate of A–1, B+2 means that the variable was judged to be 0.1 magnitude fainter than A and 0.2 magnitude brighter than B. It requires both practice and experience to be able to detect intervals of 0.1, 0.2, 0.3, 0.4 and even 0.5 magnitude. Steps larger than 0.5 magnitude are subject to large errors. The Argelander step method, e.g. A(2)V, V(3)B, is more like the fractional method, in that the steps are not necessarily tenths of a magnitude. It is used where the magnitudes of the comparisons are unknown, and arbitrary values may be established from a series of observations. If the variable is not visible, a negative estimate is

Table 14.2. Frequency of observation: selected variables

Class and type	One observation every...
Eruptive	
FU (FU Orionis)	7 nights
GCAS (Gamma Cassiopeiæ)	7 nights
INT (T Tauri star)	night
RCB (R Coronæ Borealis)	night – 3 nights
SDOR (S Doradûs star)	7 nights
Pulsating	
CW (W Virginis)	1 hour – 10 nights
DCEP (Delta Cepheid star)	0.5 night – 2 nights
L (red, irregular)	7 – 10 nights
M (Mira star)	5 – 7 nights
RR (RR Lyræ)	half–hour – night
RV (RV Tauri)	2 – 5 – 10 nights
SR (semi-regular)	7 – 10 nights
Cataclysmic	
SN (supernova)	night
N (nova)	hour – night
NR (recurrent nova)	night
NL (nova-like)	night
UG (dwarf nova)	min – hour – night
ZAND (Z Andromedæ)	night
AM (AM Herculis)	hour – night
Eclipsing	
E	10 min – night

recorded by noting the faintest comparison which is positively seen in the form [Z (the symbol '[' means 'fainter than', while ']' means 'brighter than').

Some charts show comparisons and their magnitudes by a number; e.g 35 indicates a magnitude 3.5 star, while 100 would be a comparison of magnitude 10.0. Reduced magnitudes are rounded to one decimal place (rounding .05 up to .1, and .04 down to .0).

Photography

Few UK amateurs continue a programme of variable star photography – the climate is a limiting factor. However, the astrophotographer with a non-automatic camera, or one which may be switched to manual control, using fine grain film (for example K2415) in a rural (less light-polluted) site, may find much enjoyment and interest in 'capturing' variable stars. Short exposures under these conditions with a mounted camera, a 50-mm $f/2$ lens and fast film, say 400 ISO, may reveal stars of magnitude 7 or 8 over a sky area measuring 27° by 39°. Equipment such as this could form the basis of a useful and inexpensive nova search patrol, and can also monitor the brighter variables. Note that duplicate exposures are essential in nova-searching, since a flaw on the film may appear star-like and cause confusion when checking the film against a previously exposed 'standard'.

A 35-mm single-lens reflex camera (with interchangeable lenses), or a rangefinder camera with 'B' or 'T' exposure settings, may be utilised on a simple, equatorially guided, device like the home-made Haig (or 'Scotch') mount. Alternatively, the camera could be mounted piggyback on to an accurately driven, equatorially mounted telescope, or on to a camera platform designed for the purpose. A 135-mm lens of $f/2.8$, giving a field size of 10° by 14°, and driven or guided for a 2 to 3-minute exposure, may show stars to magnitude 11 or better.

Prime focus photography, using either small refractors, larger guided reflectors or catadioptric systems (Schmidt–Cassegrain), allows fainter magnitudes to be achieved. For example, stars to magnitude 13 may be seen on 35-mm 400 ISO film, with a 6-minute exposure through a 100-mm $f/5$ refractor. Using a 250-mm aperture, some astrophotographers have reached to magnitude 16 and fainter with home-built instruments designed around robustly engineered mountings. Specialists who have the time, experience and facilities will do their own film processing, but for the less ambitious, commercially processed film may be used with success. The limiting magnitude of a photographic system depends on the lens/film combination (and in particular the film's grain size), the exposure time and the brightness of the sky background.

For a given instrumental aperture, fainter magnitudes than those visible optically will be detected by the image building process of photographic emulsions. The measuring of magnitudes of stars from negatives or printed enlargements is done either by comparison to a standard field, using the same exposure, or by other methods which include simple eye estimates, image diameter or density instruments, photometers, and microdensitometers. Ordinary eye estimates from photographs or negatives will have errors similar to those of visual techniques, while a microdensitometer measurement may be no more than four or five times more accurate.

A photographic programme might include the variable star types R Coronæ Borealis (RCB), UV Ceti flare star (UV), the red irregular (L), long period Mira (M), supernova (SN), nova (N), recurrent nova (NR) and the eclipsing class (E). Stars suspected of variability, and variables in star clusters, would also be worthwhile objects of photographic study. Amateur photographic sky patrols have resulted in many novæ and new variables being discovered in recent years.

The visual observer with a potential discovery or an important observation could do no better than confirm it with photographic evidence. This technique of observation also offers a straightforward process for making variable star charts and atlases.

Electronic Devices

Instead of the eye or a photographic emulsion to collect light, photoelectric photometry (PEP) of variable stars may be used. Here, a photomultiplier does the work by converting starlight into an electric current which is calibrated to a stellar magnitude. Typical accuracy of PEP is 0.01 magnitude or less; it is far better than visual estimation, visual photometry or photographic detection methods. A 250-mm reflecting telescope is an average size for PEP, but the equipment must be rigidly mounted and smoothly guided for tracking the variable and the comparison star. The measuring devices may use a chart recorder, a simple meter, or a microcomputer. The construction of an automatic photoelectric telescope (APT), where observations are recorded robotically, is not beyond the advanced amateur with a creative attitude and the necessary skills. The level of research available using an APT is probably comparable to that of a professional observatory. Most types of variable shown in the *General Catalogue* become accessible, given the correct aperture; work on eclipsing binaries, Cepheids and RR Lyræ types, and finding accurate times of minima and maxima, would be worthwhile. Magnitude measurements of some of the 15,000 suspected variables would confirm whether they are indeed variable or not.

The charge coupled device (CCD) has only recently

become available to advanced amateurs, who have used it to detect very deep magnitudes of 18 and fainter. The great sensitivity of the CCD reduces instrumental aperture roughly by a factor of 10, but the optical design and mechanical engineering has to be of high quality. Supernovæ and quasi-stellar objects (QSO) are regarded by some observers as high priority subjects for study using CCD technology.

References

Star Atlases

Papadopoulos C and Scovil C E, *True Visual Magnitude Photographic Star Atlas*, Vols 1–3, Pergamon, 1980
Scovil C E, *The AAVSO Variable Star Atlas* (2nd edn), Sky Publishing Corporation, 1990
Tirion W *et al.*, *Uranometria*, Vols 1–2, Sky Publishing Corporation, 1987–1988

Books

Covington M A, *Astrophotography For The Amateur*, Cambridge University Press, 1991
Hall D S and Genet R M, *Photoelectric Photometry Of Variable Stars: A Practical Guide For The Small Observatory*, IAPPP, 1982
Hoffmeister C *et al.*, *Variable Stars*, Springer-Verlag, 1985
Isles J E, *Webb Society Deep-Sky Observer's Handbook; Vol. 8, Variable Stars*, (Ed. Glyn Jones), Enslow Publ., 1990
Percy J (Ed.), *The Study Of Variable Stars Using Small Telescopes*, Cambridge University Press, 1986
Percy J R, Mattei J A and Sterken C (Eds), *Variable Star Research: An International Perspective*, Cambridge University Press, 1992
Sidgwick J B, *Observational Astronomy For Amateurs*, (3rd edn, ed. Gamble), 272–302, Faber and Faber, 1971

Magazines

Astronomy Now, London
Practical Astronomy, London
Sky & Telescope, Cambridge, Mass., USA
The Astronomer, Basingstoke, Hampshire

Reference

Kholopov P N (Ed.), *General Catalogue Of Variable Stars* (4th edn), Vols 1–4, Nauka, Moscow, 1985–1987
Kukarkin B V *et al.*, *New Catalogue Of Suspected Variable Stars*, Nauka, Moscow, 1982

Organisations

American Association Of Variable Star Observers (AAVSO), 25 Birch Street, Cambridge, MA 02138, USA

British Astronomical Association, Variable Star Section (BAA VSS), Burlington House, Piccadilly, London W1V 9AG

International Amateur–Professional Photoelectric Photometry (IAPPP), Rolling Ridge Observatory, 3621 Ridge Parkway, Erie, PE 16510, USA

Chapter 15

Supernovæ

R. W. Arbour

The Search for Supernovæ

Most novæ are discovered by amateurs, a large proportion of comets are found by amateurs, many minor planets and variable stars have recently been discovered by amateurs – and yet the number of supernovæ (SNe) discovered by amateurs is still very small. Why should this be so? Unlike the objects mentioned above, whose discovery position cannot be predicted, SNe occur in or near the amorphous glow of a galaxy, and therefore the search area is limited to a few arc minutes. Is it because they are rare? No; professionals regularly find SNe that could have been found with amateur equipment. Is it because amateurs are not interested? No; there are probably many more amateurs searching for SNe than for comets and novæ.

Professional Patrols

Undoubtedly, two very productive telescopes in the discovery of SNe have been the 48-inch and 18-inch Schmidt cameras on Mount Palomar. The 18-inch was designed specifically for searching for SNe, while the 48-inch has also been involved in such work and has made many 'serendipitous' discoveries. With these instruments, Zwicky searched for and discovered some 122 SNe, but his total is rapidly being approached by several other observers. The present leaders are:[1] Charles Kowal (83), Jean Mueller (71), Pollas (66), Wild (48), McNaught (41), Lovas (37), Antezana (36), Wischnjewsky (35), Humason (31) and Robert Evans, the leading amateur visual discoverer (27). Some of the above have since proved not to be actual SN discoveries, while some were co-discoveries. Several were serendipitous, and others have simply used other's observational plates; Jean Mueller has discovered 71 SNe while making plates for the new Palomar Sky Survey. When a new plate is made, it is compared with one taken 35 years earlier and it is relatively easy to detect either a new SN or one that might have been missed on an earlier plate. Most amateur SN searchers are envious of having a library of master plates, each covering 6°, taken with a 48-inch Schmidt camera.

At one time there were eleven nations participating in an SN search programme; although many have gradually been phased out, there has been a recent resurgence of interest, and some very interesting projects have evolved. Computer technology has been applied to search methods which have led to completely automated patrols.[2] Telescopes can be programmed to slew to a number of galaxies; a CCD camera then makes an exposure of each lasting only a few seconds and automatically compares the image with one taken previously. There is no human interference, and several hundred galaxies can be checked each night. Automated search programmes, such as the one operated by Sterling Colgate,[3] have been in operation for a number of years. It is surprising that the Berkeley automated system, which can check some 600 galaxies down to ~17 m_v (magnitude visual) on a clear winter night, has taken several years to achieve a handful of discoveries, while the Revd Robert Evans, an Australian amateur, has been successful using a much smaller telescope visually and operating the telescope manually.

Amateur Discoveries

Successful amateur SN searchers are relatively rare. The first was G. Romano, an Italian amateur from Padua, who found 1961H in NGC 4564 in 1961, followed by Jack Bennett with 1968L in the bright galaxy M83, found while comet searching. Eleven years later, Gus Johnson found 1979L in M100 which attained 11.6 m_v. In Japan,

Kiyomi Akazaki searches photographically for SNe and so far has found two; one he shared with a Russian professional and both with the Revd Robert Evans. Other amateurs to have been successful are Wayne Johnson, co-discoverer of SN 1991T in NGC 4527, and Dana Patchick who found SN 1978L in NGC 2336. More recently, Francisco Garcia, an amateur from Spain, found 1993J at 12 m_v in M81. Figure 15.1, a and b, shows Garcia's supernova as its brightness increases.

Undoubtedly, the most prolific amateur discoverer of SNe of all time is the Revd Robert Evans. This preacher from Hazlebrook, New South Wales, in Australia is a living legend among amateur astronomers, and it is doubtful if his total of 27 visual SNe discoveries will ever be surpassed. His technique is to observe as regularly as possible, and as many as several hundred galaxies each night. Unlike other amateur discoveries, Evans has found many at the limit of his instrumentation. No doubt a crucial part of his success is due to being very familiar with the appearance of literally hundreds of galaxy fields, which enables him to examine each galaxy in as short a time as possible. On one occasion he is reported to have patrolled some 570 galaxies during an observation lasting ten hours.[4]

Organised Amateur Patrols

Many attempts have been made to set up organised patrols between a number of amateurs, and many of these have been unsuccessful. Unlike a nova patrol, you cannot divide up the sky into a number of galaxies. The individual observer must be left to decide which galaxies he will observe, even though they may already be well covered. Early organised amateur patrols floundered under this régime; light pollution, accessibility and instrumentation dictate which galaxies will be searched. A far better approach is for each country to have its own amateur clearing house for suspect SNe. In Britain, this is the UK Supernova Search Programme, which acts for such organisations as the British Astronomical Association and is coordinated by Guy Hurst. The programme's American counterpart, Sunsearch,[5] is coordinated by Wayne Johnson in California. Similar groups exist in Spain and Australia. If a suspect is found, it can be rapidly checked by a number of experienced amateurs.

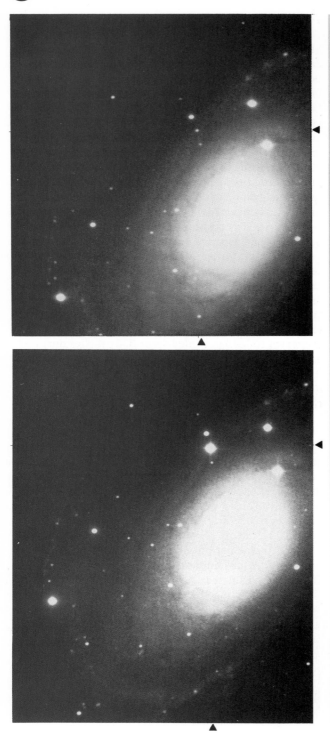

Figure 15.1a (top) and **b** (bottom). The supernova 1993J, discovered by Francisco Garcia in the galaxy M81, increasing in brightness.

A Chance in a Lifetime

Several attempts have been made to determine statistically the frequency at which a single galaxy might produce an SN. Estimates range from one every 30 years to one every 300 years. The current 'guesstimate' by the University of California at Berkeley[6] is one SN per 30 years in a galaxy such as our own. This is based on a relatively small sample. Of 20 SNe discovered by Berkeley in five years, 18 were in late-type Sc spirals, which constituted 45% of the galaxies searched. A figure of one SN per hundred years is probably more realistic if all galaxy types are included. Even allowing for several SNe which may have been missed, owing to such factors as the Solar system's position and obscuring dust, only one SN has been found in our own galaxy in the last 300 years or so.

In order to be successful, the number of galaxies searched, and the frequency of search, are vital. Galaxy morphological type is also an important factor. More SNe have been found in Sc type galaxies than any other. Out of a total of 651 SNe discovered by 1988, 132 were found in Sc type galaxies.[7] These galaxies appear to be the most productive, but the figures are misleading; it is easier to detect an SN in an Sc type galaxy, especially if it is face-on. It may well be that elliptical galaxies are more productive. Figure 15.2 shows a large, face-on, Sc-type galaxy in which, so far, four SNe have been found.

Ellipticals, and galaxies which have more pronounced

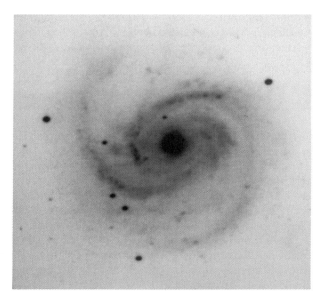

Figure 15.2 A large, face-on, Sc-type galaxy.

Figure 15.3 An Sb galaxy. Galaxies of this type are often overexposed during photography, and this can lead to a supernova being missed.

nuclear regions, are normally over-exposed in professional photographs (see, for example, Figure 15.3), and this may have helped many SNe to go undetected. This is particularly so with deep plates from Schmidt cameras, where the images of galaxies are often totally burnt out. Another factor that has probably influenced SNe statistics is the preference of searchers, particularly amateurs, to select galaxy types where SNe are more easily found. Those inclined at glancing angles or edge-on have probably been avoided, as well as those with denser nuclear regions. Not until SNe totals are much higher, and unbiased searches are conducted on all morphological types, will true production rates be realised.

Methods

There are three main techniques which can be employed in the amateur's search for SNe. These are visual, photographic and CCD imaging, and they are discussed here in order of success of detection and ease of use, and inversely proportional to their complexity, cost and difficulty of use. Contrary to popular belief, success is not proportional to financial outlay; indeed, only one SN has been found using an amateur CCD camera. To date, visual methods employing Newtonian telescopes have enjoyed the most success. The reasons for this are outlined later in this chapter.

Whichever method is chosen, the basic principle is to check all the stars which appear in the neighbourhood of a galaxy, out to two galaxy radii, and to compare them with observations made at an earlier date. For the visual observer, this previous observation will be a drawing, for the photographer a photographic print, and for the CCD user an image on a VDU or paper hard copy. A master photograph or CCD image should show stars at least half to one magnitude fainter than the intended search limit. Occasionally, the seeing or transparency will render stars visible that are normally below the limit of observation, giving rise to potential false alarms. It is essential to have master images available, to obviate wasting time with false suspects.

The master images or drawings should be obtained using the same telescope as that used for searching. Future observations will then have the same characteristics: scale, resolution, approximate stellar magnitude representation and, most importantly, no missing stellar images. Much wasted time can result from using master charts from other sources, time which could be more profitably spent searching additional galaxies. It is very common for charts to have omissions or placement errors. If they do not match the user's telescopic impression, they can cause much confusion. Although the writer prefers to make his own, this is not to say that third-party charts are a complete waste of time; the writer has various sets,[8] but they should be used only when a suspect has been found that is not on the user's master image. They will also prove invaluable in determining the suspect's brightness, field orientation and position relative to the parent galaxy.

Once an accurate idea is held of where and how bright the suspect is, immediate confirmation should be sought. The observer *must not* immediately phone, fax or telex the Central Bureau for Astronomical Telegrams, or other professional astronomical institutions! Confirmation must come from a second source, such as one of the national search teams mentioned earlier, or an amateur experienced in the observation of galaxies or SNe.

Prior to this, every attempt must be made to try to eliminate the suspect from being an SN. It might actually be a known variable star, and can be eliminated by recourse to a variable star catalogue such as the GCVS. Minor planets can also be confused with SNe, and can be eliminated with the aid of computer program lists of minor planets which may be in the field being studied.

Alternatively, minor planets will exhibit motion relative to close field stars over a period of from 30 minutes to an hour. If the suspect shows obvious motion and it is not listed in a current minor planet list, then this could be a discovery of its own, worthy of further investigation. Other sources of error include HII regions which would usually be below the limit of visibility but are rendered visible by better than normal conditions.

As the observer gains experience, false alarms become less frequent. Alerting the professional astronomical community with false alarms quickly reduces the observer's credibility, and he may even be ignored when he does eventually make a real discovery. False alarms are not limited to amateurs; professionals often blunder in this fashion. If the national clearing house gives confirmation of a 'find', they can provide the most painless route to alerting the professional astronomical community.

Visual Searching

As Robert Evans has so ably shown,[9] a visual search with a simple Newtonian is more than adequate; however, one must be prepared to search many galaxies per session to achieve success. But which galaxies should the amateur observe, and how often? The nearer, and therefore apparently larger and brighter, galaxies generally produce brighter SNe. If fainter galaxies are patrolled, then a suspect will have to be nearer maximum brightness, where it will be easier to detect. An SN will be close to maximum brightness for two to four weeks, suggesting that such galaxies should be searched every two to four weeks for the chance of catching an SN in outburst. Observing the brighter galaxies more frequently may yield an SN several days before maximum, when it may be as much as two magnitudes fainter. To catch an SN in its pre-maximum stage is invaluable.[10] If the professionals can be alerted while the SN is on the rise, valuable information on the nature of SNe will undoubtedly be gained.

We have learnt that the Sc spirals are the most prolific, and consequently the most obvious targets; therefore more amateurs will be searching them. Those galaxies where the nuclear regions are larger and more prominent, such as NGC 2841 (type Sb) and NGC 3115 (S0), are sites where visual searching has an advantage over photography. It is in the nuclear regions of such galaxies that SNe can easily be burnt out by photographic over-exposure, while visually they can often be seen to stand out against the nebulosity.

Photographic Searching

Most professional discoveries of SNe have been made using photography. It is reasonable, therefore, to expect this to be the most fruitful technique for the amateur, but only a small minority of amateur discoveries have been made in this way. The reason is one of lower coverage; the techniques involved in the photographic process reduce the number of observations, and probably deter observers. While it may take 30 seconds to check a galaxy visually, to take an exposure lasting even a few minutes may require 30 minutes to focus the camera, find and centre a guide star and make the actual exposure. On top of this, two exposures must be made to obviate possible flaws in the photographic emulsion that could be confused for an SN. There is also the added time required to process, dry and examine the negatives.

The writer has improved his coverage considerably by employing a precision drive system[11] to obviate guiding, with all its associated complexities. Two-minute exposures and a fast emulsion then make it possible to cover over 30 double fields in a standard five-hour session.

If only one or two galaxies can be photographed on each frame, the visual observer will have a considerable advantage in coverage. The astrophotographer uses a Schmidt camera centred on clusters of galaxies. There is a popular belief that the Schmidt will record fainter stars more quickly than will a Newtonian of the same aperture and focal ratio; this is simply not true. The Newtonian will, in fact, eventually reach a fainter stellar limit, since the Schmidt will reach the sky fog limit much sooner.

With photography, there is more chance of finding pre-maximum SNe, or those which have gone undetected and are well past maximum light. Because the average SN peaks at between 16–17 m_v photographers can search fainter galaxies than can visual observers.

The choice between photographic emulsions is a balance between sensitivity for short exposures with longer focal lengths and hence fainter limits, or grain size and higher signal-to-noise ratio for wider fields and hence longer exposures.

CCD Searching

The photographic searcher must be blessed with considerable patience if he is going to succeed, but this pales into insignificance compared with the problems faced by those who use a CCD camera. These devices are several times more sensitive than photographic emulsions and have greater dynamic range and a wider spectral response, but what does this mean in terms of our quest for SNe?

What used to take between five and ten minutes to record using a very fast photographic emulsion can now be achieved in a matter of a few tens of seconds with a CCD camera, without recourse to darkroom or messy chemicals. Results are immediately displayed on a monitor, and can be 'blinked' against a master image taken earlier. An interloper is then instantly obvious to the scrutineer. Image intensity values can be re-scaled to enhance or subdue certain areas, allowing the user to 'see' into overexposed nuclear regions. With the aid of a computer, the brightness of an SN can be accurately measured and its offset coordinates from the galaxy's nucleus determined. The effects of light pollution are less important using CCDs, and can be dramatically reduced by simple subtraction techniques. CCDs are much more versatile than photographic emulsions. Why, then, should such powerful weaponry be less successful at SN searching?

Once again it is a matter of galaxy coverage. Currently, the most common device available to the amateur has an active area which is no more than 2.5 mm square. Used with an instrument of 80-inch (200-mm) focal length, the field covered is only about 4 arc minutes wide, equivalent to an area only 140th that of a 35-mm frame. It is obvious that trying to centre a galaxy can be very time consuming. 'Viewing' with a 14-inch computer monitor can be likened to finding a galaxy using a power of ×700! The writer's CCD employs a device measuring 6.4 × 4.5 mm, but this is still equivalent to a power of ×30. A useful ploy is to centre the galaxy in the main telescope and change to a much higher power for more accurate centring, but the CCD must be exactly coincident with the telescope's optical axis. Quick and accurate centring of galaxies is the key to higher productivity.

Other surprises await the CCD user. Higher constraints are placed on focusing and tracking than are required for photography, and theoretically, CCDs have lower resolution than photographic emulsions. In practice, however, seeing reduces the resolution of a photographic emulsion such as Kodak Technical Pan 2415 to that of the writer's CCD, which has a pixel size of $12.7 \times 16\ \mu$.

Amateur Automated Searches

After many years of failure, professional automated searches are becoming more successful. The speed at which a vast number of galaxies can be acquired, imaged and checked to faint limits is their key to successful searching. With powerful computer systems now available to the amateur with engineering experience, it is quite possible to emulate professional automated patrols. The writer built such a system in the mid 1980s, but it was only mildly effective. Such systems are used by amateur variable star observers, but many of these employ crude centring techniques where the requirements are to slew continuously over very small arcs. If the same systems are employed for SN searching, cumulative errors soon negate their usefulness. If automatic checking is required, then the complexity of the software, and the necessary computing power, will involve a reduction in the number of galaxies searched.

The recent introduction of digital setting circles has revolutionised astronomy for those who previously had to endure 'star hopping'. These devices can contain large catalogues of galaxies, such as the NGC, so instead of taking charts or a set of coordinates to the telescope, the user chooses a galaxy number and merely guides the telescope until the digital display reads zero. These devices ensure that the required object can be found in the field of a low power eyepiece, and they are ideal for the visual or photographic observer. They will not, however, guarantee the object to be in the field of a CCD.

Another innovation is the recent introduction of first generation commercial computer-controlled telescopes. Like the digital setting circles, these are still of insufficient accuracy to place a galaxy centrally on a CCD device, but when they do, the amateur discovery rate will, no doubt, rapidly increase. Let us hope that by then light pollution will have slowed down enough so that we can still search for SNe without requiring an expensive, mountain-top observatory.

References

1   Marsden Dr B, Private communication
2   Kahn R N, 'Desperately Seeking Supernovæ', *Sky & Telescope* 73, June 1987, p594
3   Reitmeyer W L, 'Astronomy in New Mexico', *Sky & Telescope* 45, Jan 1973, p20
4   Liller W, *The Cambridge Guide to Astronomical Discovery*, Cambridge University Press, 1992, p72
5   Johnson W, 'Restructuring of Sunsearch', *Webb Soc. Deep Sky Observer* 3, Oct 1993, p3
6   'News notes: Supernova Census', *Sky & Telescope* 84, Aug 1992, p129
7   Barbon R, Cappellarno E and Terrato M, *The Asiago Supernova Catalogue*, July 1988
8   Thompson G D and Bryan J, *Supernova Search Charts and Handbook*, Cambridge University Press, 1989
9   Liller W, op. cit., p66
10 Thompson G D and Bryan J, op. cit., p59
11 Arbour R W, 'Deep-Sky Photography Without Guiding', *Sky & Telescope* 78, Nov 1989, p538

Chapter 16

Deep Sky Observing

Bernard C. Abrams

Deep sky observing is arguably the most popular branch of amateur observational astronomy. The blend of intensity and delicacy seen in the objects studied is unrivalled elsewhere, and there is plenty of scope for research should the beauty of it all prove to be insufficient motivation. However, it is not an easy 'starter' area and, in keeping with the target audience of this book, basic knowledge of observational astronomy – instrumentation and terminology as well as the sky – has been assumed, together with access to charts and catalogues. Some underpinning ideas have been included to set the context where appropriate.

Instruments: An Introduction

While the subject of instrumentation is dealt with in depth throughout this chapter, a brief word of introduction is needed. In contemplating observation of the deep sky, it is important to realise that there is no substitute for aperture. Buy or make the largest aperture telescope ($f/4$ to $f/7$) you can afford, and do not sacrifice too much in terms of optical quality, in spite of advice to the contrary which appears from time to time.

If you plan to record images photographically or electronically, go for the best quality mount and drives you can afford, then be prepared to adjust and customise for optimum results. Indeed, if visual observation is not your main interest, put the quality of mount and drives first, since the biggest and best astrographic telescope will be useless if its mount cannot track a target and if drive corrections cannot be made quickly and accurately. Do not be put off by stories of the 'essential' nature of off-axis guiding. A good 3 or 4-inch refractor will serve well as a guide telescope if securely mounted. Equally critical to successful photography is the choice of a larger minor axis flat and low profile camera mount, preferably with helical focusing. The following equations give information concerning the flat minor axis (M), mirror diameter (D), focal length (F) and the distance between flat and film plane (S), when the full 43-mm diagonal of a 35-mm film frame is evenly illuminated (to reduce vignetting):

$$M = 43 + S(D - 43)/F$$

or

$$S = F(M - 43)/(D - 43)$$

Finally, invest in the largest aperture rich field finder you can get. A 50-mm finder is barely adequate; something closer to 80 or 90 mm will be much more useful. And don't expect setting circles to save you if you don't know the sky well. There is no substitute for experience, and most local astronomical societies have plenty of that.

Polar Alignment

For ease of observing, a driven telescope is preferable and, in pursuit of deep sky photographs, almost essential (though some amazing feats of hand-guiding have been accomplished). Beyond visual sighting on Polaris, the 'drift' technique is a reliable way of improving polar alignment. In carrying out this process, drift in Right Ascension should be ignored. First, select a star which is within 15° or so of the eastern horizon and close to the celestial equator. If the star is followed at high power (a magnification of ×5 per inch of focal length is advisable, reflecting the guiding situation) and it drifts north, the polar axis is pointing too high. Following fine adjustment – unfortunately many mounts do not possess the necessary mechanisms – select a star which lies near the celestial equator and on the meridian. Again, follow the star and monitor drift in Declination only. If the star drifts north, the polar axis is pointing West of North. Following each refinement to the alignment, increasing times can be used to monitor drift. Some indication can then be gained of the amount of time a mount can be left to itself during observation. Photographic methods of achieving this, such as the King test, are beyond the scope of this book.

Deep Sky Objects

For the purpose of this chapter, deep sky objects have been grouped together under the following headings:

- Dark nebulæ
- Emission nebulæ
- Stars (limiting magnitude)
- Open clusters
- Global clusters
- Planetary nebulæ and supernova remnants
- Galaxies

In each sub-section, some underpinning background information has been included where it is helpful in illuminating an aspect of observational relevance. No treatment such as this can hope to be exhaustive, and the examples listed are a small and personal selection from a vast catalogue of targets (further information can be found in publications such as *Norton's Sky Atlas* and *Uranometria*). Both visual and photographic/electronic observational techniques are discussed as appropriate.

Dark Nebulæ

Dark nebulæ, consisting mostly of cold molecular hydrogen, are revealed only by the presence of luminous material nearby or along the line of sight. Photographs of the Milky Way star clouds show numerous ghostly silhouettes, many adjacent to bright emission nebulæ. When looking at images of M42, the Great Nebula in Orion (see Figure 16.1), it is difficult to appreciate that its true form is hidden from sight by a much larger yet unseen component. The dark material cutting into the nebula's central region (the Fish Mouth) is part of a large molecular cloud which obscures much of the emission region and robs it of greater symmetry. An equally well-known sight, the Trifid Nebula (M20), is trisected

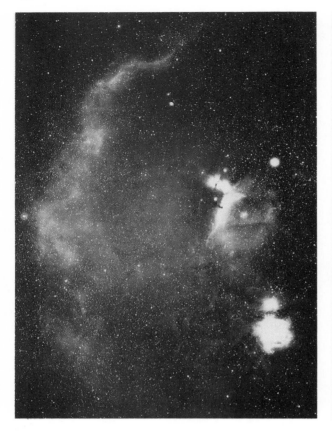

Figure 16.1 The Great Nebula in Orion, much of which is hidden by dark matter from a large molecular cloud.

Tamron 180-mm f/2.5 with H–X filter, 70-min. exposure

by dark lanes which give the nebula its name. Perhaps the most famous pair of dark blue nebulæ, which together typify the association with emission nebulæ and star fields, are B(Barnard)33 (the Horsehead nebula in Orion) and B142/3 in Aquila. Many discoveries – and catalogue numbers of dark nebulæ – are due to Barnard, the well-known and distinguished Milky Way astronomer.

The degree of obscuration, which usually ranges from 1 to 5 magnitudes or more, can be estimated by comparing the intensity ratio – predicted : observed – between optical spectral lines (which are attenuated efficiently) and associated radio features (which are not). In spite of the low density and low temperature, some exotic molecules have been formed in dark blue nebulæ, and have been identified using infra-red and microwave observations, two of the techniques capable of penetrating these galactic clouds. While the conditions used to synthesise ethanol on Earth involve temperatures and pressures which result in 10^{20} molecules per cubic centimetre of gas, it is surprising to find such compounds being formed by chemical reactions in conditions when the mere hundred or so molecules in the same volume rarely collide.

Many of the more subtle nebulæ catalogued by Barnard are best revealed by long exposure wide-field photography. A dark site is needed, where a rich field telescope or camera lens will show the objects to best effect. The southern hemisphere offers such panoramas, some of which can be seen from mid-northern latitudes. A rich collection of star fields, dark clouds and emission regions can be found in the constellations of Scutum and Sagittarius. To begin, try 20–30-second stationary camera exposures using a 50-mm lens and fast (400 to 3200 ASA/ISO) film. Use a sturdy tripod and a cable to release the shutter (set to B). Colour transparencies are least susceptible to commercial processing errors if you cannot develop and print your own work. A good tip is to include one daylight scene at the start of the film, to prevent cutting errors. For good, dark skies, limit your exposure to the square of the focal ratio you use; in light polluted skies, half this time is a safer limit (this rule of thumb can be extended to other types of deep sky photography). To obtain a narrower field of view, try using a moderate telephoto lens – 135-mm to 500-mm – on a Scotch (Haig) mount, or piggybacked on a driven equatorial. Table 16.1 shows the angular field of view obtained on a 35-mm film frame, using various telephoto lenses.

Targets suitable for the visual observer are mostly those objects associated with bright nebulæ. As well as the Orion and Trifid complexes, the Flame Nebula (NGC 2024) and the Lagoon Nebula (M8) are fine examples. Many of the isolated regions of dark matter in these nebulæ, known as Bok globules, are believed to be the

Table 16.1. 35-mm frame field for a given focal length

Lens focal length (mm)	35-mm frame field (°)
50	28 × 42
135	10 × 15
200	7 × 10.5
300	4.5 × 7
400	3.5 × 5
500	2.8 × 4.2

site of star formation. Indeed, this process is one trigger which turns a dark cloud (HI region) into an emission nebula (HII region). The areas mentioned above are large stellar nurseries; even modest dark clouds contain several hundred solar masses. External galaxies, mainly spirals and irregulars, have similar dark clouds which we can observe from an external vantage point. In the case of spirals, the dark nebulæ are largely confined to the plane of the star-forming regions in the spiral arms.

Emission Nebulæ

When star formation has produced sufficiently luminous objects, a dark interstellar cloud will be excited by radiation from stars and protostars, and will glow with colours characteristic of the elements present within it. On film, the red spectral line of hydrogen is prominent, while at the eyepiece such emission nebulæ frequently appear green (when colour can be detected at all), due to a less prominent oxygen feature which happens to occur at a wavelength to which the retina is particularly sensitive. Blue coloration detected around emission nebulæ and young stars is due to scattering of blue light, and is not an emission phenomenon – such areas are frequently referred to as reflection nebulæ.

Visually there are many impressive emission nebulæ. Most deep sky observers begin with M42, the Great Nebula in Orion, and M8, the Lagoon Nebula in Sagittarius, both naked eye objects under the skies – while with even a telescope of moderate aperture the detail visible can be overwhelming. One can then move on to objects such as M20 (the Trifid Nebula in Sagittarius), M78 (Orion), and, more difficult, IC 5146 (the Cocoon nebula in Cygnus) and NGC 7635 (the Bubble Nebula in Cassiopeia). To assist with sketches of any extended deep sky object, try mentally dividing the field of view into quadrants, as this will help with positional detail. Averted vision will bring fainter parts of a nebula within reach. Sketching on to tracing paper taped over a red torch face (if flat enough) will prevent loss of dark adaptation, and if the tracing paper has a circle divided into quadrants then transcription may be further eased. Some observers with DIY skills have taken this idea one stage further by inserting a transparent plastic disc, etched with a grid system, into the focal plane of the eyepiece. This arrangement will provide the basics of a coordinate frame of reference when illuminated from the side by a red LED in the manner of a guiding eyepiece.

The nature of the light from emission nebulæ allows a high-technology solution to the problem of light pollution. Since the spectral lines characteristic of emission nebulæ are mostly situated away from the lines of light pollution (sodium, and some mercury), interference filters can be used to remove the light pollution selectively. Some attenuation of nebular light does occur, but the gain in signal-to-noise ratio can bring previously invisible objects into view. For visual observers the O III filters can produce excellent results, particularly with planetary nebulæ (see p. 245), while photographers will find the less specific deep sky filters beneficial.

Choosing an eyepiece for deep sky observing can be as much a matter of cost as suitability. I have obtained excellent views with relatively inexpensive Plossls and Erfles. The wider field of view obtained with Naglers can be spectacular, but more light loss is bound to occur with the larger multi-element designs.

In scanning the literature and catalogues for references to nebulæ, the magnitude values quoted can be confusing and misleading. Integrated magnitudes do not give as true an indication of visibility as visual surface brightness magnitudes, which are rarely given. To gain an impression of the visibility (or otherwise) of an object prior to observation, it may be better to ask somebody who has actually seen it than to consult magnitude lists.

Photographically, emission nebulæ provide some stunning views of the heavens. Exposures of only a few minutes' duration can reveal impressive amounts of nebulosity. As a compromise between light-grasp, cost and manageability, the traditional 10 or 12-inch $f/5$ to $f/6$ reflector has much to recommend it. With a well-aligned equatorial, skilful guiding and an appropriate choice of film and processing, views can be obtained which will rival professional photographs. Black and white films such as Tri-X and T-Max were firm favourites for many years, particularly when combined with an energetic developer such as D-19 or MWP-2. Modern high-speed colour emulsions have much to offer, particularly E6 chemistry films (transparencies) for the reasons outlined earlier. For optimum results, photographers use a variety of techniques to improve the performance of their films. Cold cameras and hypering can be used to good effect, as can post-exposure techniques such as intensification, dodging and stacking multiple negatives in the enlarger (mono-chrome).

(Stars) Limiting Magnitude

This relatively short section on limiting magnitudes reflects the degree of uncertainty inherent in discussions of this kind. There are so many variables operating that comparisons are almost impossible to make; sky conditions, collimation and quality of the optics, eyepiece effects, accuracy of focus, experience and optical health of the observer – all affect visual observations; and to these factors can be added the type of camera, choice of film, processing method, consistency in use of special techniques (e.g. hypering), exposure duration and guiding accuracy when discussing photographic observations. However, rather than admit defeat in an area which arouses much interest and controversy, I will attempt to do that most dangerous of things and generalise. Over many years as an active deep sky observer and astrophotographer, I have derived empirical expressions which have provided some predictive idea of the limiting magnitude to be expected, when working with good equipment used to best advantage under 'normal' conditions.

The formulæ are (M = limiting magnitude, D = aperture in inches, E = exposure in minutes):

Visual limit: $M = 9.5 + 5.0 \log D$
Photographic limit: $M = 9.0 + 5.0 \log D + 2.2 \log E$

Table 16.2 shows these formulæ for a range of apertures.

The second part of Table 16.2 has been constructed to show the exposure times needed to reach a limiting magnitude of approximately 15 with a variety of apertures. In comparison with a range of results obtained by colleagues with different instrumentation and techniques, in varying conditions, the fit has been found to be least good at the edge of the ranges quoted. The fit is best when comparing fast reflectors, and (photographically) when comparing the use of monochrome emulsions rated at between 400 and 1000 ASA/ISO.

Open Clusters

In Charles Messier's famous *Catalogue* of nebulous objects, open clusters rank third – behind galaxies and

globular clusters – in abundance; in addition, many
more can be found in the NGC and other lists. Although
superseded by Trümpler's classification, the Shapley
scheme is still used. Very loose, irregular groupings are
given a 'c' label, with more condensed, richer clusters
ranging from 'd' through to 'g'. The category 'a' (aster-
ism) can be used to designate very poor groupings
which may not be true associations. Open clusters are
excellent targets for the binocular observer or small
telescope user. Many of them lie just beyond the naked
eye threshold and are revealed and resolved by the most
modest of optical aids. Lying within the star-forming
regions of the Galaxy's spiral arms, in the galactic
plane, open clusters are irregular collections of up to a
thousand or so stars which are bound together gravita-
tionally.

While only a few hundred such objects are visible,
there may be over 100,000 clusters in the Galaxy, mostly
hidden from view by cold dark matter in the spiral
arms. The majority are young objects, ranging in age
from 10 million to 1000 million years. Consequently, few
red giants are found in open clusters, as most member

Table 16.2: Limiting magnitude for a range of apertures

Visual Limit

D (inches)	M (limiting magnitude)
4	12.5
6	13.4
8	13.9
10	14.5
12	14.9
16	15.5

Additional values for larger aperture instruments can be calculated using the formula, but these will become increasingly unreliable as extrapolation proceeds

Photographic Limit

D (inches)	E (minutes)	M (limiting magnitude)
4	20	14.9
6	10	15.1
8	5	15.1
10	3	15.0
12	2	15.1
16	1	15.0

stars are either on the main sequence or evolving towards it. Although they are born together and remain so for some time, gravitational forces from external objects and internal stellar encounters gradually disperse the member stars. Very rich groups, such as the Wild Duck cluster (M11 in Scutum) and those situated out of the plane of the Galaxy (M67 in Cancer) survive the longest. Careful observations and long exposure photography will reveal a relative lack of fainter stars in such objects, a circumstance which is due in part to the dynamic escape of low mass members. The core of a cluster is all that we can detect visually. For example, the Pleiades (M45 in Taurus) is often listed as having an apparent diameter of one degree, yet member stars – located by virtue of their proper motion – have been found across 10° of the sky. For the Hyades cluster, also in Taurus, this scatter is over 30°, equal in area to about 60 full Moons.

Near the very centre of an open cluster, a 'hard binary' which formed early in the cluster's life involving two of its most massive members, may be responsible for preventing core collapse. Interactions with other cluster members lead to this binary becoming tightly bound, and give energy to other stars, thus preventing collapse. Indeed, some clusters (such as M35 in Gemini) give the impression of having relatively vacuous centres. Some areas of sky contain notable 'clusters of clusters', such as the Perseus Double Cluster, the Scutum area, and the Auriga trio (M36, M37 and M38).

Analyses of such groups, where distance values make nearby objects truly close (and not a line-of-sight effect) suggest that cluster formation may be initiated in more than one location by the same trigger event, possibly a nearby supernova explosion. It is difficult to avoid the feeling of awe which open clusters inspire when viewed with a large aperture at low power. There are 'useful' avenues of work open to suitably equipped and dedicated observers of open clusters – monitoring flare stars in the Pleiades, for example – but it is more often for æsthetic reasons that open clusters in the winter and summer Milky Way are observed. In addition to those mentioned above, some of the more attractive clusters include M34 (Perseus) and M39 (Cygnus) – both good binocular targets – together with M42 (Canis Major), M52 and M103 (both Cassiopeia), and the powdery NGC 1245, easily found at a position 3° south-west of Alpha Persei.

Globular Clusters

Globular clusters are large, spheroidal groups of between 100,000 and 1,000,000 stars. They contain some of the oldest stars in the Galaxy, typically 10^{10} years of age. With a diameter usually close to 50 pc, they are ten times larger than their open cluster cousins, but since they are also some ten or more times as distant, their apparent diameters are comparable or smaller. This makes them poor targets for binoculars or small telescopes. One or two notable exceptions exist, however, such as M13 in Hercules (northern hemisphere) and Omega Centauri and 47 Tucanæ (southern hemisphere). Other objects which stand out include M3 (in Canes Venatici), M5 (Serpens), M10, M12 and M14 (in Ophiuchus), M15 (Pegasus), M22 (Sagittarius), and M92 in Hercules. For the very keen, the 'intergalactic tramp' cluster NGC 2419 (in Lynx) and the Palomar globulars will together provide many hours of entertainment. A noteworthy feature of our Galaxy's globular clusters concerns their distribution in a roughly spherical arrangement around the Galaxy's centre. There are relatively few other stars (the so-called high velocity stars) in the halo, with the vast majority of the Galaxy's stellar members located in the spiral arms of the galactic disc. Halo globular clusters could be a consequence of the more spherical shape adopted by the Galaxy in the early stages of collapse and star formation. Globular clusters are relatively stable in comparison to their open cluster cousins.

Observations of globular clusters show that, in spite of the obvious similarities, each object has distinguishing features. Distance has an effect on angular size and resolution, with the degree of condensation or compression also relevant to the view obtained. Catalogues and charts often classify globulars on a 12-point scale using Roman numerals. Clusters labelled category I are the most compressed and concentrated, and are unlikely to show many individual stars, while those in category XII are loosely compressed and easier to resolve. Observers intending to study globular clusters should use large apertures, as recommended for general deep sky observing, but longer focal lengths (and/or higher magnification eyepieces) will be useful in resolving some of the more compressed objects. Autumn and spring skies provide the best collections, since the obscuring matter of the Milky Way is less of a problem; Hercules,

Sagittarius and Ophiuchus offer many excellent targets which illustrate the range of morphology displayed by these objects.

Working with different apertures – or with a range of exposures in photography – can produce very different impressions of globular clusters, more so than other objects. Small apertures and short exposures reveal the distribution of luminous giants, while larger telescopes or moderate exposures will reach stars evolving along the horizontal branch towards the giant status. Only the largest visual telescopes, and the deepest exposures, will penetrate down to those (fainter) dwarf stars left on the main sequence in such clusters.

Planetary Nebulæ and Supernova Remnants

These two classes of deep sky object represent late or final stages in the evolution of stars. As observational targets they provide challenging and intriguing views. Since the vast majority of the stars we can observe lie in the plane of the Galaxy, this is the place to look for most planetary nebulæ and supernova remnants. Planetary nebulæ are infamously misnamed. Appearing as small discs, some of the more symmetrical of these do resemble some of the planets viewed at low magnification, but this is where the connection ends. Planetaries are isolated, gaseous nebulæ consisting of a shell of gas and dust that has been ejected in the recent past (astronomically speaking) from a central star. This star is, or soon will be, a white dwarf, and is hot (50,000 to 300,000 °Celsius), but less luminous than galactic O-stars.

Doppler shift measurements show that the shell is expanding outwards at velocities of around 25 km/s, and the visible lifetime is typically 20,000 years. The energy output near the central star is sufficient to ionise many of the elemental gases contained in the shell, and many planetaries contain significant amounts of ionised atomic oxygen, nitrogen and hydrogen. Containing about a solar mass or so of material, the shell of a planetary may be roughly a light year or so across. The line emission nature of the radiation from planetaries, involving 'pure' colours, allows us to make use of high-technology visual and photographic accessories for identification and observation. Any selection of a dozen or so well-known planetaries would show the varying degree of symmetry possessed by such objects, and also the wide range in angular size.

Some of the largest and faintest of planetaries have only recently been identified on wide-field sky survey plates, while some of the smallest are difficult to distinguish from field stars. Broadly speaking, the former category are unrealistic targets under UK skies, though there are ways of improving your chances with both. O III filters offer the best opportunity to view planetaries under light-polluted skies. If both direct and averted vision observing fail, try 'blinking' by passing the filter in and out between eye and eyepiece, with a hood over the head to exclude all unwanted light. Spotting a small planetary against a rich Milky Way star field can be assisted by using a medium power and a diffraction grating (say 80 to 200 lines per mm). When the grating is placed between eye and ocular the light from stars in the field – consisting of a wider spectrum of colours – is spread out and so considerably dimmed, while the relatively monochromatic light from an almost stellar planetary will remain bright in the primary band. Well-known objects include the small but impressive Ring Nebula (M57) in Lyra, the famous Dumb-bell nebula (M27) in Vulpecula, M76 (the mini Dumb-bell in Perseus), M97 (the Owl Nebula in Ursa Major), NGC 6781 (Aquila) and NGC 7009 (the Saturn Nebula in Aquarius).

In looking at photographs of these planetaries, remember that different emulsions give different responses. With some of the more symmetrical shell objects, the presence of red (hydrogen) in the outer regions and green (oxygen) closer to the central star reflects the lower ionising power of radiation as the distance from the white dwarf increases. It is unlikely that the eye will see such shades of colour in most planetaries using moderate apertures, and green alone will most likely appear in larger instruments.

Supernova remnants (SNRs) tend to be more irregular and to reflect the violence of their birth. Stars which are more than about 8 solar masses may not be able to shed enough material during the later stages of their evolution to get below the size limit for white dwarfs. Above a limiting mass, collapse beyond a white dwarf will take place, producing a supernova explosion and an SNR around it.

Depending on the nature of the progenitor, the SNR may be filled (a plerion, with a neutron star or pulsar) or not. At present the majority view is that all single,

massive stars will evolve to undergo supernova explosions, producing neutron star remnants or black holes. There are two broad categories of SN event: Type I (involving a star in a binary system, which leaves no stellar remnant); and Type II (involving single, highly-evolved giants, in the process described above).

The Crab Nebula (M1 in Taurus) is possibly the most famous SNR and probably the most observed object in the skies (see Figure 16.2). We are fortunate in our galactic geography, since the focused beams of radiation from the Crab neutron star sweep across the Earth and so allow us a study of the beaming mechanism. The event which produced the Crab was recorded in July of the year AD 1054 by Chinese astronomers as a new, bright 'guest' star near one horn of the Bull. Several centuries later John Bevis's telescope – trained on the same patch of sky – showed the Crab Nebula. Shortly afterwards Charles Messier observed this object while locating the comet of 1758 between the Bull's horns. The Crab Nebula was given first place in Messier's *Catalogue* of nebulous objects. Compared to planetaries, SNRs may contain around 10 solar masses of material (or more), and expand at greater velocities – up to 30,000 km/s initially. Photographs of the filamentary structure of the Crab nebula taken some tens of years apart reveal the continuation of this expansion.

Figure 16.2 The Crab Nebula in Taurus, a supernova remnant and probably the most observed object in the skies.

Some large, very old SNRs have been found (the visible lifetime of these is estimated at around 100,000 years). Some SNRs have only ever been observed at radio or X-ray wavelengths. The largest SNR is not visible from northern hemisphere skies. The Gum Nebula, named after its discoverer, Australian astronomer Colin Gum, was initially thought to be hydrogen made visible due to ionisation by two bright stars nearby. Later, the theory arose that the ionisation had been caused by a burst of radiation from a supernova many thousands of years ago; the Vela pulsar lies close to the centre of the nebula. The Gum Nebula is so large that it is best imaged using a mosaic of wide-field photographs; like the Hyades, it spans 30° of sky. Another elderly remnant is the Veil Nebula in Cygnus (NGC 6960/6992/6995), thought to be the SNR of an event which occurred over 100,000 years ago. As with the symmetrical planetaries such as the Ring Nebula, in many cases we see arcs or circles rather than shells, since we are looking through a greater thickness of material at the edge of a shell than in the middle. The Veil can be glimpsed visually – I have once managed this feat unintentionally, while setting up a photograph using a low power 80-mm finder – on a very dark night. Again, quality filters can help improve the situation, if this is something you would like to achieve.

Galaxies

Before stellar evolution reached its present stage, galactic evolution ideas were tentative and depended not so much on physical arguments (e.g. galactic dynamics) but on intuitive notions based on observations of galaxy shapes.

The first such theory was due to Hubble, and relates to the 'Tuning Fork' classification of galaxies as irregular, elliptical, spiral or barred spiral. Hubble's theory was that galaxies began as large, spherical objects (ellipticals) which then flattened due to rotation, becoming increasingly eccentric, before transforming into a tightly wound spiral. Further flattening and loss of material from the nucleus produced the more open spirals, with almost no nucleus and extensive spiral structure. Eventually a final stage was reached where all structure disappeared and an irregular galaxy remained. Interestingly, Shapley proposed that exactly the reverse sequence operates. What the Hubble classification may

Figure 16.3
Galaxy M31 offers a
wealth of detail in a
large Dobsonian, or
on film.

Takahashi 8-inch f/4,
1 hr, TP2415 (H)

be showing us is not so much an evolutionary sequence
as a conservationary one; how good certain galaxy
types are at conserving their star-forming material.
Elliptical galaxies seem to have exhausted their supply
with a burst of early, rapid star formation, while the
irregulars have been much more conservative. Present
theories of galaxy formation suggest that the initial collapse of, and subsequent star formation in, a protogalaxy are influenced by (among other factors) the original rotational velocity of the protogalaxy. In this model,
protogalaxies with moderate to little angular momen-
tum will allow rapid early star formation and produce
ellipticals, while more rapid rotation will generate
galaxies with more steady star formation, such as the
spirals.

Spring and autumn skies are the best for observing
galaxies, since then they are less obscured by the Milky
Way. Deep sky targets on offer include a large number of
galaxy groups, stretching from the northern galaxies of

Ursa Major down to Leo and Virgo which are accessible from both hemispheres. Many observers in the northern hemisphere will have glimpsed M31 with the naked eye, and gone on to enjoy more revealing views using binoculars or telescope, which will also show the satellite galaxies M32 and M110. As the other prominent member of the Local group, M31 offers a wealth of detail in a large Dobsonian or on film (see Figure 16.3). There are several Cepheid variables which can be monitored, and spotting the global clusters of M31 is a sport which attracts many followers. Other favourites include M33 (Triangulum), M74 (Pisces) and M51, the Whirlpool Galaxy in Canes Venatici, which shows distortions in the spiral due to the gravitational interaction with its neighbour, a smaller but more massive elliptical. When observing barred spiral galaxies, such as M109 (Ursa Major), or NGC 7479 in Pegasus, it is sobering to recall that astronomers have not yet explained satisfactorily the mechanism behind the origin and maintenance of the bar structure.

Some of the most impressive sights are those provided by edge-on spirals. The cold, dark matter in the spiral arm appears as a thin, dark line bisecting a needle of light. Popular examples are NGC 4565 in Coma Berenices, NGC 891 (in Andromeda; see Figure 16.4) and M104 (the Sombrero Galaxy in Virgo). In addition to observing detail in galactic structure, ranging from the visibility of arms in a spiral structure to individual HII regions or OB associations, systematic galaxy obser-

Figure 16.4 A spiral galaxy seen edge-on: NGC 891 in Andromeda.

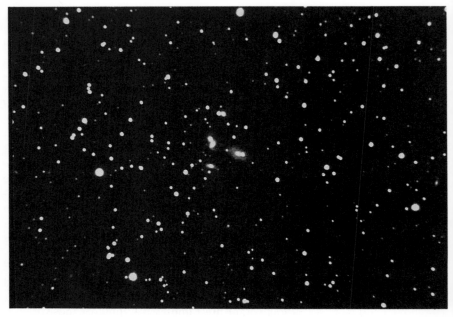

vation may produce a supernova discovery, the subject of Chapter 15. For the observer requiring a more difficult challenge, there are many faint galaxy groups which can be accessed with today's large amateur telescopes, from Stephan's Quintet in Pegasus (see Figure 16.5) to the Abell clusters of galaxies. One of the Abell clusters worth trying for, given the instrumentation and the inclination, is A2151 in Hercules.

Figure 16.5
Stephan's Quintet, in Pegasus, presents more of a challenge to the amateur observer.

f/6, 80 min. exposure

Equally challenging, and potentially very fruitful as a line of research, are the enigmatic active galactic nuclei (AGN), a category of objects which includes quasars (QSOs), BL Lacertæ objects, and Seyfert galaxies. Some AGNs are the subject of continuous interest, particularly 3C 272 in Virgo; others are less well-known but of similar importance, such as Mk (Markarian) 205, situated near NGC 4319, Mk 421 (near the bright field star 51 Ursæ Majoris), OJ 287, a BL Lacertæ object in Cancer, and the Seyfert galaxy M77 in Cetus. To participate in an observing programme involving such objects will require standardisation of procedure, photographic observations being best suited to B-filtration and blue-sensitive emulsions.

Because of the red sensitivity of many of the CCDs available commercially in the amateur market, these devices have not yet had a large impact in an area where their quantum efficiency and digitised images make them an otherwise ideal choice.

In looking back over several decades of amateur deep sky observing, certain events stand out as the benchmarks of their day. Early attempts at astrophotography revolutionised deep sky observing, with Walter Pennell and then Ron Arbour leading the way in the UK. As astrophotography grew in popularity, so did visual observing with the advent of the Dobsonian light bucket. Following commercial Schmidts and SCTs, CCDs have in turn contributed to the joy and scientific gain derived from deep sky observing. Looking to the future, the growth of the CCD use is certain to continue, with larger, cheaper and more panchromatic chips emerging.

Drive correction may be obsolete shortly, with computer processing of digital images removing drive errors and other flaws from collected data. To an extent, this is the present, not the future, albeit on a small scale. Further pro–am collaboration will assist in this evolutionary process, with the division between the two categories of activity blurring beyond distinction. What will change is the fascination inherent in the view of our universe provided by deep sky observing, whether this is undertaken in pursuit of pleasure, knowledge, or (for the well balanced observer?) both. There can be no more rewarding branch of observational amateur astronomy.

References

Jones K G, *Messier's Nebulæ and Star Clusters*, Faber (1990)

Norton A P, *Norton's Star Atlas*. (This is invaluable, but the older editions are better than the revised *Norton 2000*. In my opinion the maps in the new edition are less clear and, as they are printed across the 'gutter' of the book, they are difficult to use for plotting. PM)

Tirion W, *Cambridge Star Atlas 2000*, Cambridge University Press (1991)

Many valuable handbooks are published by the Webb Society.

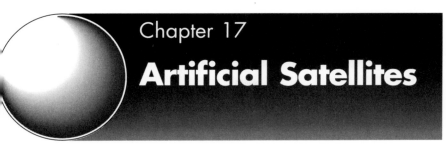

Chapter 17
Artificial Satellites

Howard Miles

Background

Most, if not all, amateur astronomers will have noticed quite frequently the passage across the sky of artificial satellites, the brightness of which can vary from about magnitude -2 down to the limits of naked eye visibility and, with optical aid, to the limiting value of the instrument being used. From the launch of *Sputnik 1* in 1957 until the end of 1991, there were about 3400 launches, and for some time the annual launch rate was about 120. Roughly two thirds of the launches were from what was the Soviet Union. The launch rate has fallen recently, there being only 95 in 1992 and only 61 in the first nine months of 1993. These figures refer to launches only, and do not reflect the actual number of objects put into orbit. Many of the rockets which put an instrumented probe into orbit are far brighter than the actual probe. Many of the objects seen with the naked eye are such rockets. The term 'satellite' is generally used, however, to include all man-made objects seen crossing the sky.

There are two main conditions which must be met before a satellite can be seen. Virtually all satellites shine only by reflected sunlight, and therefore the object must be illuminated by the Sun. It will not be seen if it is within the Earth's shadow. The other important condition is that the observer's sky must be dark, or nearly so. Generally, this means that the Sun must be more than about 8° below the observer's horizon. The degree of darkness needed for sighting, however, depends on the brightness of the satellite to be observed. These two conditions limit the time at which satellites can be seen. There are normally two visibility periods, one just after sunset and the other just before sunrise. The duration of these visibility periods depends on the height of the satellite above the Earth's surface, its orbital inclination, the time of the year and the latitude of the observer. At midwinter, the duration can be much less than an hour, but during the summer months at middle latitudes the two visibility periods merge and satellites can be seen throughout the night.

It is quite difficult to predict accurately the magnitude of a satellite for a particular transit, because there are so many variables. The figures quoted on prediction sheets can therefore only be used as a general guide. The absolute magnitude, often quoted on predictions, is the satellite's half-phase magnitude at a range of 1000 km, when seen at an altitude of 90°. The numerical value depends on the nature and curvature of the reflecting surface; flat surfaces can produce very bright reflections. If the satellite is rotating, the magnitude of the object can vary considerably and, in some cases, the satellite can flash regularly or irregularly. In other cases, one can observe a gradual variation of the light, in a regular or irregular manner, depending on the way the object is tumbling.

Table 17.1 gives the change in magnitude for distances differing from the standard distance of 1000 km. Some predictions give magnitudes which allow for the range. The change in the magnitude with altitude is negligible for altitudes greater than about 70°, but for lower altitudes the fall-off in magnitude is noticeable. Table 17.2 gives the apparent decrease in magnitude for altitudes of 70° and under.

Satellite Nomenclature

Many satellite and space probes are given popular names and these are used extensively in literature con-

Table 17.1. Change in magnitude for various ranges (based on an absolute value at 1000 km)

Range (km)	200	400	600	800	1000	1200	1600	2000	3000	4000	8000
Increase in magnitude	3.4	2.0	1.0	0.5	0	−0.4	−1.0	−1.5	−2.4	−3.0	−3.5

cerned with the on-board experiments. These names are also widely used by the media. The system breaks down, however, for two main reasons. It cannot cope with all the components, such as the spent rockets and nose cones, and because differing organisations sometimes use the same name for completely different satellites.

For scientific and catalogue purposes, each component of a launch is given a unique reference number. There are two main systems.

1. A system of numerical catalogue numbers (SSC), assigned by USSPACECOM. Objects 1 and 2 refer to *Sputnik 1* and its rocket, launched in 1957 and numbers are now assigned in sequence when the objects are listed in the public satellite catalogue.
2. The International Designation system, suitable for computer listings, which consists of a seven digit number which can be split into three parts. The first two digits refer to the year of the launch, the next three record in sequence the number of the launch for that year. The last two digits identify the component of the launch, the payload usually being referred to as 01 and the rocket which put it into orbit as 02. The system allows for 99 fragments to be listed.

For example, *Satcom C4* (an American communications satellite launched on 1992 August 31) has three principal components:

- The satellite itself:
  SSC number 22096
  International designation 9205701
- Delta 2 rocket (2nd stage):
  SSC number 22097
  International designation 9205702
- Satcom C4 rocket:
  SSC number 22098
  International designation 9205703

Table 17.2. Decrease in magnitude of satellites at various altitudes (compared with an absolute magnitude at altitude 90°)

Altitude	10°	20°	30°	40°	50°	60°	70°
Apparent decrease in magnitude	3.8	2.3	1.5	1.0	0.6	0.3	negligible

The International Designation Number is often given in a modified form, and this is widely used. In the above example the satellite would be referred to as 1992–57A, with the Delta 2 rocket being given as 1992–57B. This modification has the advantage of a systematic listing of all fragments of space junk associated with that launch. For example, in the case of the Russian Space Station *MIR*, 1986–17A, fragment 1986–17FJ was identified as probably the remnant of an unsuccessful attempt to deploy an inflated aluminium-covered balloon on 1991 August 15. The fragment decayed on 1991 August 29.

Prior to 1963, when the number of launches was relatively small, use was made of the Greek alphabet instead of numbers. For example, the American communications satellite *Telstar 1* was known as 1962 Alpha–Epsilon 1. But the increasing number of launches soon made this system unwieldy.

Reasons for Observing Satellites

In most branches of astronomy, an observer can contribute directly to the knowledge of a particular subject, the extent of this contribution depending on the optical instrument being used and the skill and experience of the observer. This is particularly true for lunar and planetary observations. With satellite observation, skill and experience are the dominating factors for making an observation, but the using of these observations falls generally outside the ability of the observer. The observations are forwarded to a central collection agency where they are sorted, and all the observations of a particular object are then analysed by mathematicians at research establishments. In the UK this work is currently being carried out at some universities, such as Southampton and Birmingham, but in the past the Space Division of the Royal Aircraft Establishment, now called the Royal Aerospace Establishment, has been the main centre for orbital analysis.

Analysis of the behaviour of a satellite in orbit has produced results which could not have been achieved from ground-based measurements. The first of these achievements was to obtain a very accurate value for the flattening of the Earth. Prior to the space age, the generally accepted value for this was 1/297.1 but, by analysing the orbital behaviour of the early satellites, the much more accurate value of 1/298.25 was obtained. This was quickly followed by the realisation that the two hemispheres were of different shapes – the so-called pear shaped Earth – with the northern hemisphere having a polar radius some 40 metres greater than that of the southern hemisphere. This difference was detected because it produced a variation in the distance of perigee of a satellite's orbit from the centre of the Earth. The perigee distance oscillates regularly between two values as the position of perigee moves round the orbit of the satellite.

With analysis of the orbits of more and more satellites, further refinements have taken place. For example the Earth's equator, once taken to be a circle, has been found to be elliptical with maximum values at longitudes 20°W and 160°E. Refinements to this simple picture have been made, giving a much better understanding of the real shape of the Earth. Recent developments have shown that there are variations along a particular latitude, so the Earth can be considered as being covered by a mosaic of what are termed 'tesseral harmonics', which record the variations in the shape of the geoid for the whole surface.

The study of the shape of the Earth has many important implications for the study of the Earth's interior, but orbital analysis is by no means limited to studies of the Earth's shape. Much has been learned about the upper atmosphere, its structure, distribution and movements. The way solar activity influences the atmospheric pressure at a given height has been deduced by the effects of atmospheric drag on the behaviour of a whole range of satellites.

In the current economic climate of the UK, severe restrictions have been placed on the prediction service, and it has been found impossible to issue predictions for a large number of satellites. A selected list of objects has been chosen for their usefulness in subsequent research, and predictions are issued for these. It must be pointed out that, in all such cases, it is the orbital behaviour of the object that is being investigated, this having no connection with any on-board experiment.

The main areas of research which lead to a satellite's inclusion on the selected list are:

- Determination of air density and its variation
- Determination of atmospheric rotational speed
- Geostationary satellites
- Study of observations using lasers

- Studying behaviour when satellite nears its decay
- Analysis of flash periods and magnitude
- Analysis of lunisolar resonance
- Evaluation of odd zonal harmonics in the Earth's gravitational field
- Evaluation of 13th, 14th and 15th order harmonics in the Earth's gravitational field at resonance, and 29th order harmonics at 29:2 resonance.

In addition, some of the brighter satellites are included in the list for the purpose of training observers and others for amateur analysis of their orbits.

For readers who are not conversant with the term *harmonics*, the Earth's shape can be considered to be a perfect sphere which is modified by a series of more complicated shapes, or harmonics. The second harmonic produces a tendency towards an elliptical shape, while the third harmonic, tending towards a triangular shape, allows for the so-called pear shape of the Earth. The higher harmonics produce shapes corresponding to the number of high points in the shape. By superimposing these harmonics and adjusting the numerical values of their coefficients it is possible to produce a shape which corresponds closely with the true shape of the Earth. The mathematics used to calculate these coefficients is beyond the scope of this book, but any reader interested in this work is referred to King-Hele (1987).[1]

Behaviour of a Satellite in Orbit

If the Earth were a perfect sphere; if there were no atmosphere; if the Sun and Moon did not exist; and if the Earth did not have a magnetic field, then a satellite once placed into a particular orbit would remain in that orbit without change moving round the Earth in accordance with Kepler's Laws of Motion. It would then be possible to produce predictions years in advance and, generally speaking, there would be little point in tracking satellites. However, none of the above conditions apply, so the actual path of a satellite deviates from the pure orbit by amounts which vary according to the height of the satellite, its orbital eccentricity and the inclination of the orbital plane to the equator.

A satellite will experience the most atmospheric drag when it is at perigee, where the atmosphere will be densest. The loss of energy occurring during a perigee passage prevents the satellite from reaching the previous apogee height. The net effect is that the apogee falls fairly quickly, whereas the perigee changes very little; the eccentricity gradually changes, but the rate at which this takes place depends on the initial height of the orbit. Obviously the eccentricity of low satellites will change far more quickly than that of higher ones. As the eccentricity approaches zero, i.e. the orbit becomes roughly circular, the satellite will experience high drag throughout its orbit. This will cause the object to spiral slowly inwards, eventually to be burned up in the atmosphere to produce a bright fireball. A satellite having a perigee of lower than 120 km has a life of only a few orbits, but one with a perigee greater than 2000 km experiences very little drag and has a very long life.

Solar activity has a big effect on the density of the upper atmosphere. The pressure at a given height can alter in quite a short space of time by a factor of over 100. This obviously alters the drag forces, and produces complex orbital behaviour. Solar radiation pressure can produce very large and complex perturbations, especially on satellites with large area/mass ratios.

The fact that the Earth is non-spherical produces two main effects on the orbit. If our knowledge of the shape of the Earth is accurate, these effects, unlike the causes described above, can be predicted.

The Earth's equatorial bulge causes the orbital plane of the satellite to regress, i.e. to rotate about the spin axis of the Earth like the wobbling of a spinning top, and causes the major axis of the ellipse to rotate in the plane of the ellipse. The rates of these rotations depend on the orbital inclination and the average height of the satellite above the Earth's surface. The values of these effects, in degrees per day, can be calculated from the formulæ:

Rate of regression (X) =
$$0.964(R/(R + h))^{3.5}(1 - e^2)^{-2}\cos i*$$

Rotation of the major axis =
$$4.982(R/(R + h))^{3.5}(1 - e^2)^{-2}(5\cos^2 i - 1)*$$

where R = equatorial radius of the Earth
  h = average height of the satellite above the Earth's surface
  e = eccentricity
  i = inclination of the orbital plane.

---

* Unless the satellite is in a highly eccentric orbit, the term $(1 - e^2)^{-2}$ can be considered as being equal to 1.

These effects, together with the fact that a satellite normally does not make an integral number of revolutions in exactly 24 hours, cause the satellite to travel over different parts of the Earth's surface each day. The daily westward movement of the orbital plane is given by

$$X + 1 + n/4$$

where X = regression (see above)

1 = the average value for the westward movement of the stars due to the Earth's motion round the Sun.

n = difference in time in minutes between 24 hours and the time for an integral number of revolutions completed in 24 hours.

Observation

Predicting the Path of a Satellite

Some of the perturbations mentioned in the previous section can be predicted accurately, but those due to the state of the atmosphere can only be estimated. However, a sudden change, say in solar activity, can play havoc with predictions. As far as the general observer is concerned, the problem is to obtain from the predictions the track across the sky as seen by an observer located at a known position on the Earth's surface. It must be emphasised that this will be a predicted track and not necessarily the actual one. The main purpose of making satellite observations is therefore to report the difference between the observed and calculated positions. All the observer needs to do is to report the observed position of the satellite, and the time it was there.

There are four main ways of obtaining predictions:

- Using 'Look Angle' predictions obtained from a prediction centre, specifically prepared for a known observing location
- Using the orbital elements to generate the 'Look Angle' data, in conjunction with a computer program. Without the use of a computer the calculation would be far too long and impractical
- Using a graphical technique involving tables of computed values and predictions based on the time and

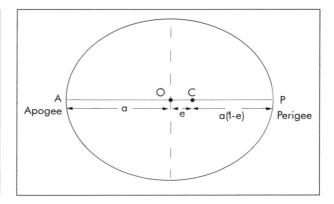

position of the ascending node, height and inclination of the orbit
* 'Do-it-yourself' predictions based on observations made during the satellite's previous transits.

In each method it is necessary to obtain or derive all or some of the orbital elements. Quite often, for example with newly launched objects, the elements may not be available, but enough limited information may be gleaned from the media for rough predictions to be calculated – hence the value of the graphical techniques mentioned above.

To obtain accurate predictions it is necessary to know the numerical values of the orbital elements. The basic elements are:

* The length of the semi-major axis of the ellipse (a)
* The eccentricity of the ellipse (e).

(These two quantities define the size and shape of the orbit; see Figure 17.1.)

* The orbital inclination (i), i.e. the angle between the orbital plane of the satellite and the equator
* The position of the ascending node (Ω), i.e. the position where the satellite crosses the equator in a northbound direction, measured by the angle between the ascending node and the First Point of Aries.

(These two elements define the orientation of the ellipse in space.)

* The argument of perigee (ω), i.e. the direction of

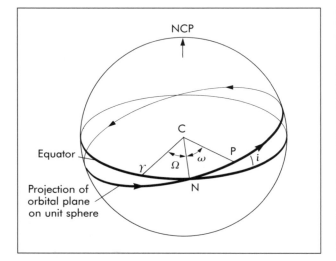

Figure 17.2 The elements which define the orientation of an orbit in space:

C centre of the Earth
N ascending node
P perigee
ϒ first point of Aries
i inclination of the orbital plane to the Equator.

the position of perigee within the orbital plane defined as the angle between the position of perigee and the ascending node measured in the direction of motion (see Figure 17.2)

- The time at which the satellite was at perigee (T).

In satellite observation, these elements have been modified to give the information in a more suitable way. The 'Two-line Elements' issued by prediction centres contain, along with other data, the following information which is directly useful for calculating 'Look Angle' information, i.e. the predicted track across the sky as seen from a particular location on Earth.

- Satellite identification.
- The date and time at which the elements used had the values used in calculating the predictions, i.e. the epoch.*
- The number of revolutions per day (n).
- Half the rate of change of n per day (n/2).
- The inclination of the orbit (i).
- The Right Ascension of the ascending node (Ω).
- The eccentricity (e).

---

* The epoch is given in the form of a number consisting of five digits prior to the decimal point and eight after the point. The first two give the year and the other three before the decimal point the day of the year. The digits after the point define the fraction of the day. For example, 93331.52551415 gives an epoch 1993 November 27 at 12hr 36min 44sec.

- The argument of perigee (ω).
- The mean anomaly (M). This gives the position of the satellite in its orbit in a useful form. For a better understanding of the mathematics of this quantity, the reader is referred to any book which deals with orbital calculations (e.g. Roy 1978[2]).

Computer programs are available for handling these elements to produce a series of predicted locations for a satellite as it crosses the sky, as seen by an observer at a given position on the ground. If the program is suitably adapted, it will also give information on the position of the Earth's shadow and whether or not the satellite is visible.[3]

Before computers were generally available, the prediction centre provided the 'Look Angle' directly. In fact this service is still available in a modified form, but the data is specific for six locations in the UK. The majority of UK observers are not more than 150 km from one of these locations, so the errors introduced are mostly quite small and in some cases can be ignored. These predictions provide the azimuth of rising and setting and also the azimuth and altitude for the position when the satellite is highest in the sky. Also provided are the azimuths for the points where the satellite enters (or leaves) the Earth's shadow.

For many years the most common form of prediction was known as the 'equator crossing' form. This form provided data on the time and position of the ascending node, together with a subsidiary table allowing positions away from the node to be obtained. By using simple graphical techniques a series of Look Angles could be found.[4]

Equipment for Observing Satellites

Although there are many naked-eye satellites, some of which are exceptionally bright, far more accurate observations can be made if an optical aid is used. There are two basic requirements for making useful observations. The first is to determine the position of the satellite with reference to the star background. The second requirement is to record as accurately as possible the time at which it was at that position.

In general terms, for obtaining a precise position, the

requirements of any optical aid are light gathering ability and a wide field of view (a minimum of about 5°), so that in addition to being able to locate fainter satellites, fainter stars can be used to determine the position. Magnification is of secondary importance. It has been found that 7×50 or 11×80 binoculars are the most useful aids, although because of their weight, some observers prefer to mount the latter on a tripod. Elbow telescopes, such as those used on gun sights, have been used successfully by many observers in the past, although they have the disadvantage of looking into the instrument at right angles to the line of the satellites. Wide field telescopes can be used but, except for specialised purposes such as observing geostationary satellites, ordinary astronomical telescopes have very limited use, because of their narrow field of view and lack of manoeuverability. Theodolites can be used, provided that they are mounted very accurately. They provide data in the form of altitude and azimuth, but the method is not recommended for the inexperienced observer.

During the last few years, digital stop watches have become widely available. Because of their improved accuracy over the analogue types it is strongly recommended that these should be used for timing. Associated with the stop watch, it is necessary to have access to a standard time signal. These are generally in the form of radio time signals, and observers should be able to pick up signals from at least one of the transmitting stations. Alternatively, the telephone 'Speaking Clock' can be used in the UK as a standard. The only other requirement is access to star maps. Maps such as *Norton's* and the British Astronomical Association's star charts are useful for drawing in the predicted track, so that it is possible to locate the best regions for observing the object, bearing in mind the position of the Earth's shadow. However, these maps must not be used for determining the position of a satellite against the star background. For this purpose, use must be made of more accurate maps or catalogues.[5]

Making an Observation

To make an observation which will be of use to an analyst, it is necessary to record as accurately as possible the position of a satellite at a given time. Because making the fix involves simultaneously recording both the

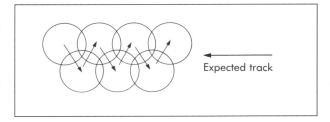

Figure 17.3
Scanning along the expected track to locate a faint satellite.

Expected track

position and the time, it is desirable to develop a systematic technique.

The first step is to plot a series of predicted positions in Right Ascension and Declination on a star map, with information regarding the position of the Earth's shadow. This will define the region of the sky where the satellite is expected to be. It is advantageous to assess the suitability of the star background along the whole of the predicted track. Some parts of this track may be more favourable for making the fix. One region of the sky may consist of a very large number of faint stars, so it must be remembered that any reference stars must be readily identifiable. To locate a faint satellite it may be necessary initially to observe a strip of sky wider than the field of view of the instrument. A technique similar to that illustrated in Figure 17.3 is suggested. In all cases the scan should be carried out against the direction of travel of the satellite, so that there is no danger of missing the satellite on a re-run.

Once the satellite is located, it is then necessary to fix its position against the star background. Ideally this occurs when the satellite passes between two identifiable stars which are oriented so that the line joining the stars is roughly at right angles to the line of travel of the satellite. The distance between the reference stars should be as small as possible. The satellite's position when it crosses the line joining the stars is estimated as a fraction of the distance between the two stars (see Figure 17.4.a). The best type of fix occurs when the satellite actually occults a star, but this happens rarely. More common is a 'grazing occultation' when the satellite and the star almost appear as a single object. This, together with a knowledge of which side of the star the satellite passed, can produce a very accurate fix. Quite often the satellite will pass just outside the two chosen stars (see Figure 17.4.b). An estimate of the distance to the nearest star can be made, but the accuracy falls off rapidly with distance from the star. Accuracy is also reduced if the line joining the stars makes an angle of

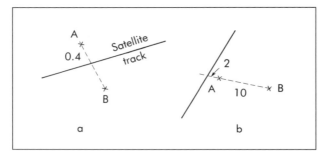

a   b

Figure 17.4a, b
Methods of fixing the
position of a satellite
with reference to the
star background.

less than 60° with the track of the satellite, so in this case
it is normally better to choose other reference stars.

It is rare that the positions of the stars relative to the
satellite are ideal, so it is often necessary to make a compromise. This is where the skill of the observer is important – a classic case of practice makes perfect.

Obtaining a time fix tends to be more difficult than
fixing the position. The techniques used depend largely
on personal preference. The watch should be checked
and rated accurately; for digital watches, however, errors
over such a short period of time will generally be very
small. The most common method is for the observer to
start the watch at the time of making the positional fix
and then stop it against the standard time signal (for a
method comparable to this, see the example in Chapter
12: Occultations, on pp. 187/8). A point worthy of note
here is that the time signal's minute marker is quite often
indicated by a longer pulse, which may come unexpectedly, so it may be advantageous to stop the watch a specific number of seconds after the minute marker.

With practice it should become possible for the
observer to make two or three timed observations during a single transit.

Reducing and Reporting an Observation

The receiving centres usually require observational data
to be presented in a standard format, and the appropriate reporting forms can be obtained from these centres.
To help the beginner, the Artificial Satellite Section of
the British Astronomical Association has produced a
simplified form upon which any reduction to be made is
carried out by the Section. The purpose of this is to
ensure that the beginner is making reasonably accurate

observations and to resolve any potential difficulties.

The satellite's position should normally be reduced to Right Ascension and Declination, using either an accurate star map or a catalogue. Under no circumstances should approximate star maps be used. In all cases the method used should be reported, as well as the epoch of the star map used. The appropriate receiving centre will provide details of the charts and catalogues available at the time. Details of the standard time signal used must be included in the report. Radio stations transmit Coordinated Universal Time (UTC), and the transmissions (except OMA) also carry a coded correction so that it is possible to convert UTC to UTL. However, all observations should be reported in UTC.[6]

Decay of Satellites

Apart from some minor differences, the phenomena associated with the decay of artificial satellites (or their fragments) and natural meteoric material are very similar. The main difference between them is that the former travel across the sky much more slowly. The fireball produced by both types of object is quite spectacular and can range in brightness from that of the planets to values exceeding that of the full Moon. Much can be learned from these events and, for the UK, the British Astronomical Association runs a Fireball Survey. Any reader witnessing such an event is requested to send in a report to the BAA at Burlington House, Piccadilly, London, W1V 9AG.

References

1   King-Hele D, *Satellite Orbits in an Atmosphere*, Blackie 1987
    This provides a mathematical formulation of the evolution of satellite orbits experiencing air drag. It is not suitable for readers without a good mathematical background.

2   *Any book on astrodynamics or elliptical orbits will give a description of the mathematics involved in calculating elliptical motion. One such book is:*
    Roy A E, *Orbital Motion*, Adam Hilger 1978

3   *A computer program for obtaining Look Data from the orbital elements has been written by Gordon E. Taylor and is known as ARTSAT. An up-to-date version of the program can be obtained from:*
    Robert Harrols, 10A Barker Avenue, Rose Heyworth Estate, Abertillery, Gwent NP3 1SE

4   *With the availability of personal computers, this method has fallen out of use. The methods of obtaining local predictions from Equator Crossing Data have been described in:*
    Miles H, *Satellite Observer's Manual*, British Astronomical Association 1973
    Miles H, *Artificial Satellite Observing*, Faber and Faber 1974
        For methods used in the early days of satellite observation, the following is of historic interest:
    Heywood J (Ed.), *Artificial Earth Satellites*, British Astronomical Association 1961

5   *Star atlases which are widely used for reducing observations include the following:*
    Atlas Coeli 1950.0, A Becvar
    Atlas Borealis 1950.0, A Becvar
    Atlas Eclipticalis 1950.0, A Becvar
    Atlas Australis 1950.0, A Becvar
        These last three cover the whole sky; the Borealis, latitudes greater than 30°N, the Eclipticalis, latitudes between 30°N and 30°S, while the Australis covers the southern hemisphere. Stars down to magnitude +10 are recorded on maps having a scale of 2 cm per degree. A transparent grid enables accurate positions of RA and Dec. to be obtained.
        New atlases with Epoch 2000.0 have recently become available, details of which can be obtained from prediction centres.

6   Seidelmann P K (Ed.), *Astronomical Almanac, Nautical Almanac Office, US Naval Observatory*, University Science Books, Mill Valley, Cal. 1992
        This gives details of the various time systems in use, and their derivatives.

Other publications of interest to satellite observers

King-Hele D, *A Tapestry of Orbits*, Cambridge 1992

An excellent description, mathematical in parts, of the work on orbital analysis carried out by Desmond King-Hele while he was on the staff of the Royal Aircraft Establishment.

King-Hele D, *Observing Earth Satellites*, Macmillan 1983

A non-mathematical review of the whole field of satellite observing.

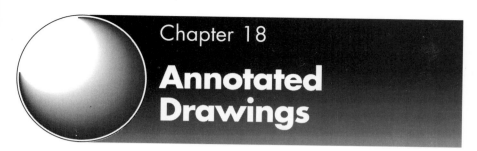

Chapter 18

Annotated Drawings

Paul Doherty

There are many ways to record what is observed through a telescope. Both conventional photography and CCD imaging are excellent methods. In the case of CCDs, results are beginning to surpass most other forms of recording. Unfortunately, these methods require expensive equipment which is often beyond the reach of most amateurs.

What we are concerned with here is recording observations by drawing what is seen through the telescope and accompanying these drawings with adequately detailed notes. This method is relatively simple, and is certainly far cheaper than photography. That is not to say, however, that drawing at the telescope is easy. Observers will encounter difficulties, but patience and practice will soon overcome these, and will lead to self discipline and a consistency of approach which is vital should you wish your observations to be of scientific value.

Equipment

Most astronomical observing is carried out in the dark, and time is usually at a premium. It is wise to ensure that all the necessary materials are contained within a single unit. The ideal is a clipboard (maximum size around A4) with an attached, battery-powered light, and a pencil fixed by string. The pencil should preferably have an eraser at one end.

The type of light to be used is worth some thought.

For faint objects it should be dim and red. While observing in the dark the eye becomes very sensitive to low luminosity, as the pupil widens gradually to accept more light. This is known as 'dark adaptation', which usually continues over a of period between 20 and 30 minutes, maybe even longer. It is a very valuable asset and should not be wasted, which can occur instantly if the eye is exposed to a bright light for no more than a fraction of a second.

For brilliant objects, such as the Moon or brighter planets, dark adaptation is not so important; in fact, the reverse can sometimes be the case. If the light directed on to your drawing paper is too faint while observing bright objects, it will be a while before you can see the paper itself, and valuable time could be lost. Here a variable light is the ideal, and the best way to achieve this is by use of a slip-on red filter cap which can be added or removed depending on the object under scrutiny. Pencil lights are a good idea, but these usually throw a shadow over the drawing area which can be a nuisance.

It is often impossible to remain seated during an observation – indeed telescope attitude usually makes this impossible. It is therefore advisable to have a set of sturdy steps available, particularly if they are fitted with an upright post to hang on to.

Drawing Materials

With these initial details sorted out, the next thing to consider is drawing materials. Drawings should be made on card of about the same thickness as the average postcard, which should be smooth but not too smooth. Some surface grain is necessary for the paper to accept pencil work easily. On the other hand, if the surface is too rough, the resulting drawing will also appear rough and less pleasing. The card needs to be of a good quality, otherwise the surface will damage easily and the paper fibres will break up, giving the finished drawing a ragged effect. Doubtless you will also wish your artistry to last for many years, so it is worth checking that the card will not deteriorate or discolour quickly.

Choice of pencil depends on your chosen subject, since no single grade of pencil will do everything you require. For ease, however, keep pencil variety to a minimum. HB and F are ideal general grades for mid-range shades, while the harder H to 3H grades are good for lighter shades such as faint planetary detail. Soft B to 2B pencils are best for heavier shadings, such as deep lunar shadows. Anything harder than 3H is likely to damage even the finest paper surfaces in any but the most experienced of hands, and will be of very limited use. At the other end of the scale, work done with pencils softer than grade B smudges very easily and is unlikely to last. It is possible to use smudging to good effect: rubbing with either a finger or a cotton bud will produce large areas of smooth, even tone. It is even worth considering shaving the softer pencils to make a fine powder, which can be spread with cotton buds to achieve greater areas of even tone easily and quickly.

Use of a sharpened eraser can also be employed to produce bright areas of spots which may be observed in the shaded regions, and to shape the shaded areas themselves. The main problem is that this type of work may be easily damaged and needs to be treated with care. Remember, also, that accuracy is paramount. It should never be compromised for the sake of a 'nice drawing'. At the same time it must be appreciated that a good draughtsman stands a better chance of making accurate representations of what is seen through the telescope.

Prepared Blanks

Each object under study will demand a separate skill, and there are aids which will make life easier. In the case of planetary work it is advisable to have prepared blanks representing the shape of the planet to be studied at the time of the observation. These will increase your chances of drawing in its correct position any feature seen, thus adding scientific value to your observation – your drawings may eventually be required for detailed analysis; who knows, a rare event may be recognised on one of them in the future.

All of the bright planets change their appearance, size and shape, and this must be accounted for. Mercury and Venus show a full range of phases, like the Moon, and the predicted percentage of their illuminated disc, as seen from Earth, should be accurately marked on a pre-prepared 2-inch diameter blank. Any deviation from the predicted phase can then be noted, and any markings seen can be drawn accurately in accordance with the disc's true shape.

Mars also shows phases, though never the full range like Mercury or Venus. Nevertheless, a 2-inch diameter blank disc should be prepared before an observation and, again, any slight phase can be accounted for.

Jupiter presents another problem. Its disc is not circular but appears to be slightly flattened at the poles. A template is valuable here, and should be in the form of a 70° ellipse with a major axis (equatorial diameter) of 2.5 inches. One master is perfectly adequate, since Jupiter does not appear to change a great deal.

For the inner planets, one blank for every 1% of illumination would be the ideal but, since this would require 50 masters, one blank for each 5% would be more convenient. You only need to prepare those between 1% to 50% illumination, turning these through 180° for the remainder of the sequence.

Saturn is the most problematic of all the planets, for the purposes of illustration. Its globe is also an elliptic spheroid, slightly more flattened than that of Jupiter, and it is of course girdled by a ring system. This system changes its tilt from 0° to 28° to our line of sight over a seven-year period, and goes through a full range of opening and closing both to the North and then to the South over a period of nearly 30 years.

It is possible to prepare your own Saturn blanks, but this is a long and involved process. Serious planetary observers will doubtless wish to join an association which can make use of their work, e.g. the BAA in the UK or ALPO in the USA. These organisations actually supply prepared blanks for the various planets, and the convenience of this alone is well worth the joining fee.

If, however, you do decide to prepare your own Saturn blanks it is advisable to make a master from which can be traced the required outline for an intended observation. One outline for each degree of ring tilt, from 1 to 28, will suffice, but they do not all need to be prepared at once – since the variation takes place over a seven-year period, the series can be built up gradually. On average the variation throughout a single apparition is usually 3°–4°, so these are the outlines you need to prepare each year. The full set will be completed in seven years, and these can be used again for the following seven years and then inverted for the remaining 15 years of the full cycle. Afterwards the whole process begins again, so the outlines can be used for as long as is required.

For the black background of your blanks it is best to use 'process black'. This is a water-based medium, giving a matt finish which photographs as a dense black.

Comets

Observers of comets will have different requirements. Here, photography scores in many ways, but visual observation has its merits. Because long exposures are needed to show the faint, outermost regions of a comet's tail, inner detail close to the nucleus is always burned out by overexposure.

These regions are particularly fascinating because it is from them that most of the larger-scale activity stems. Experienced observers will be able to see fine details close to the comet's nucleus, such as jet activity, which may even give a clue to its rotation period.

Drawings of comets are easiest if made in negative form, using pencil on white paper. For the more adventurous, powdered chalk applied with a cotton bud will give a more realistic effect. Airbrush users will find their method ideal for the rendition of comets but, as always, accuracy is important. Even though they may look impressive, inaccurate illustrations are worse than useless.

Notes

All observational drawings should be accompanied by notes which give essential information not shown on the drawings themselves. Obviously, the more details the better but, in the absence of time, notes may have to be cut short. In that case, there is certain information without which even the finest drawing would be of no value, and so must not be omitted. These essentials are:

- Date of observation
- Time of observation in Universal Time (UT = GMT beginning at midnight)
- Telescope type, size and magnification used; e.g. 12-inch (30-cm) reflector (refl) (or refractor (OG)), magnification = ×300
- Seeing conditions: use the Antoniadi scale as follows:

  1. Perfect seeing, without a quiver.
  2. Slight undulations, with moments of calm lasting several seconds.
  3. Moderate seeing, with larger air tremors.
  4. Poor seeing, constant troublesome undulations.
  5. Very bad seeing, unsuitable for anything except a rough sketch.

Drawings should form only a part of the observational record and, in any case, seeing conditions will often make a full-blown illustration impracticable. However, a rough sketch is better than no drawing at all. Should this be impossible, brief notes are better than nothing, as long as they contain the above essentials.

Intensity Estimates

If time allows, there are other notes which will add value to the observation. For planetary work, intensity estimates of any feature, dark or bright, should be made, and there are guidelines for the scales to be used in order to gain consistency between observers when a general analysis is to be made. These are as follows:

For Mercury and Venus: Use a scale of 0–5, with 0 = the brightest feature observed, 5 = the darkest.

For Mars, Jupiter and Saturn: Use a scale of 0–10, with 0 for the brightest feature and 10 for the sky background. (In the case of Saturn, 1 = the brightness of outer ring B).

A particularly useful type of observation for Mars, Saturn and Jupiter is the determination of longitude for any interesting feature seen. This is especially important for Saturn and Jupiter, which have gaseous surfaces. By making timings as features cross the exact centre of the planet's disc at an imaginary line called the Central Meridian (CM), you can work out longitudes for any feature at a given time. In this way it is possible to establish individual drift rates and therefore the speeds of atmospheric currents at various latitudes. Tables of longitude and drift rate, with time for both of the accepted systems of rotation in Jupiter's atmosphere (Systems I and II), are published by the BAA and ALPO.

Comets: Once again, comet observations require a different approach. In this case the intensity and structure of the brightest part of the object, the head or coma, is graded on a scale from 0–9 (see Table 11.5, p.173).

Additionally, the total brightness of a comet should be noted. This is not always easy since it usually involves comparing an extended object with a point source, a star. The easiest way to accomplish this is to use comparison stars around the comet and defocus the telescope until they appear to be the same size as the comet itself, and then find one of similar brightness, the mag-

nitude of which should be known. From this, an estimate of the comet's brightness can be made. If there are
no stars nearby which are comparable to the comet, and
you have no information about the magnitudes of the
stars you have used, make a scale of five estimates from
0 to 10 including the comet. For example:

Star 1:	0 (Brighter)
Star 2:	3
Comet:	4
Star 3:	7
Star 4:	10 (Fainter)

Make a sketch of star positions around the comet,
indicating those which were used for the estimates.
From this information it will be possible for an analyst
to quantify your estimate of the comet's brightness.

If a tail is present, its length in degrees of arc should
be estimated, together with its position angle (pa), the
latter being estimated in degrees again, but measured
from north through east on a scale of 0 for north, 90 for
east, 180 for south and 270 for west. To find north with
an equatorially mounted telescope is easy: simply swing
the instrument upwards on its Declination axis and the
last part of the original field to disappear from view will
be the northernmost point. Altazimuth telescopes do
not work in this way, so simply sketch the field stars and
use a star atlas to determine the location of north.
Finally, the diameter of the comet's head, or coma,
should be estimated in arc minutes.

Never record anything you are uncertain of in a sketch;
instead, make a note of your suspicion but state your
uncertainty. There will be no pleasure in looking back
on your work in future years – and no value in it for
others – if any of the information is suspect.

Astronomical recording using these techniques is
useful and can give a lot of pleasure, if attention is given
to the brief points outlined in this chapter. Once you
gain experience, the task will become easier and you
will wish to take things further.

Contributors

Bruce Hardie is the Director of the Solar Section of the British Astronomical Association.

Bob Turner is a very active solar observer, who has his observatory in Sussex and is a leading member of the BAA Solar Section.

Michael Maunder is a leading eclipse observer, and a Council Member of the BAA.

Jeremy Cook is the Director of the Lunar Section of the BAA.

Richard Baum is Director of the Mercury and Venus Section of the BAA.

Patrick Moore is a Past President of the BAA, and has been observing Mars regularly since 1934.

Terry Moseley is President of the Irish Astronomical Society, and a regular contributor to the Jupiter Section of the BAA.

Alan Heath was for many years Director of the Saturn Section of the BAA.

Andrew Hollis is Director of the Asteroids and Remote Planets Section of the BAA.

Neil Bone is Director of the Meteor Section of the BAA.

Jonathan Shanklin is Director of the Comet Section of the BAA.

Alan Wells is Occultation Coordinator of the Lunar Section of the BAA.

David Gavine is a leading member of the Aurora Section of the BAA.

Melvyn Taylor is a former Director of the Variable Star Section of the BAA.

Ron Arbour is a former Director of the Deep Sky Section of the BAA, and is currently a Member of Council.

Bernard Abrams is a former Director of the Deep Sky Section of the BAA.

Howard Miles is Director of the Artificial Satellite Section of the BAA, and is a Past President of the Association.

Paul Doherty is a leading astronomical artist whose work is known in many countries, and is often featured on television.

Index